CORVETTES

Images & Stories About
America's Great Sports Car

Harvey Goldstein

Amherst Media, Inc. ■ Buffalo, NY

Published by:
Amherst Media, Inc., P.O. Box 538, Buffalo, N.Y. 14213
www.AmherstMedia.com

Publisher: Craig Alesse
Associate Publisher: Katie Kiss
Senior Editor/Production Manager: Michelle Perkins
Editors: Barbara A. Lynch-Johnt, Beth Alesse
Editorial Assistance from: Ray Bakos, Carey A. Miller, Rebecca Rudell, Jen Sexton-Riley
Business Manager: Sarah Loder
Marketing Associate: Tonya Flickinger

ISBN-13: 978-1-68203-338-8
Library of Congress Control Number: 2018960367
Printed in The United States of America.
10 9 8 7 6 5 4 3 2 1

www.facebook.com/AmherstMediaInc
www.youtube.com/AmherstMedia
www.twitter.com/AmherstMedia
www.instagram.com/amherstmediaphotobooks

AUTHOR A BOOK WITH AMHERST MEDIA

Are you an accomplished photographer with devoted fans? Consider authoring a book with us and share your quality images and wisdom with your fans. It's a great way to build your business and brand through a high-quality, full-color printed book sold worldwide. Our experienced team makes it easy and rewarding for each book sold—no cost to you. E-mail **submissions@amherstmedia.com** today.

Contents

About the Author

Harvey Goldstein (Cr.Photog.) resides in Branford, Connecticut, and has been in the photographic industry for 45 years. Harvey and his brother Alan were the owners of Alfa Studio in Middletown, Connecticut, for 23 years.

Harvey is a Past President of the Connecticut Professional Photographers Association (CTPPA) and the Professional Photographers Association of New England, and later served as the Executive Manager for CTPPA. He served on the Bylaws, Rules, and Ethics Committee of the Professional Photographers of America (PPA) for 26 years, as well as on the Nominating Committee for the Professional Photographers of America. He was a member of PPA's Council from 1980 to 2018.

He was the editor for the Connecticut Professional Photographers Association for 32 years as well as the editor for the Professional Photogra-

phers Association of New England for 19 years. He also served as editor for the American Society of Photographers, the Professional Photographers Association of Massachusetts and the Maine Professional Photographers Association. He was the editor for *LENS* magazine, published by CPQ Professional Imaging in Cleveland, Tennessee (1999–2006); a contributing writer for *Rangefinder* magazine and *AfterCapture;* and has been an editor for Amherst Media since 2013.

Catch up with Havey at: www.facebook.com/harvey.goldstein.5.

Introduction and Acknowledgments

I have always been in awe of those wonderful, sleek specialty cars produced by the major automobile manufacturers. My earliest memory of these unique cars was my uncle's 1955 Ford Thunderbird with the porthole. I was taken with the looks of the T-Birds and the MGs that the college students drove when I was six years old—as well as the Volkswagens, for those who could not afford an MG, the Edsel (but, sadly, that did not last long), and of course, the early Chevrolet Corvette Stingray. To my dismay, when I finally got my first car, it was a Nash Rambler.

The Corvette was, and is, the definitive car for those who long for speed and beauty. Some cars are made for one or the other, but the 'Vette encompasses everything an automobile aficionado desires under him or her. While many of the Corvette owners I encountered in the production of this book were men, it is certainly not an exclusively male-dominated club.

The stories in this book are from the owners of these magnificent vehicles—and there were a few stories that I heard, but I was not allowed to share.

Needless to say, many of the untold stories came from the women and how they obtained their Corvette from a former husband or boyfriend. You can have my house, but you can't have my Corvette!

Corvette owners have a special wave when they see each other on the road. To all 'Vette owners who read and enjoy the stories in this book, "Save the wave!"

I want to thank the Corvette owners from all over the United States who contributed to this book, as well as the Corvette Club of Connecticut, the Corvette Connection in Bolton, Connecticut, and the Corvette Center in Newington, Connecticut.

A big thank you goes out to the primary photographer for this project, Glenn Curtis and his wife Marie Curtis (M.Photog., CPP) and Anastasia Fasnakis from Curtis Photography Studio (www.curtisphotographystudio.com) in Middlefield, Connecticut.

This book is dedicated to Nancy, my loving and extremely patient wife.

–Harv Goldstein

Full of Surprises
1957 Convertible, Red

Owner: Wayne

Wayne already owned a 1968 Corvette, but he had always been enamored with Corvettes with single headlights. For a few years, he searched for a 1956 or a 1957 model and finally found an affordable one 200 miles from his home. Wayne recounts the experience:

"My preference was a red car with a four-speed transmission. The one I found was a blue 1957 with automatic transmission, but I bought it anyway. After I drove the car home and began working on it, I discovered that the original color was red. Further investigation showed that the clutch pedal was pushed down and covered by the carpet. These discoveries convinced me that this was the car I was meant to buy, restore, and cherish."

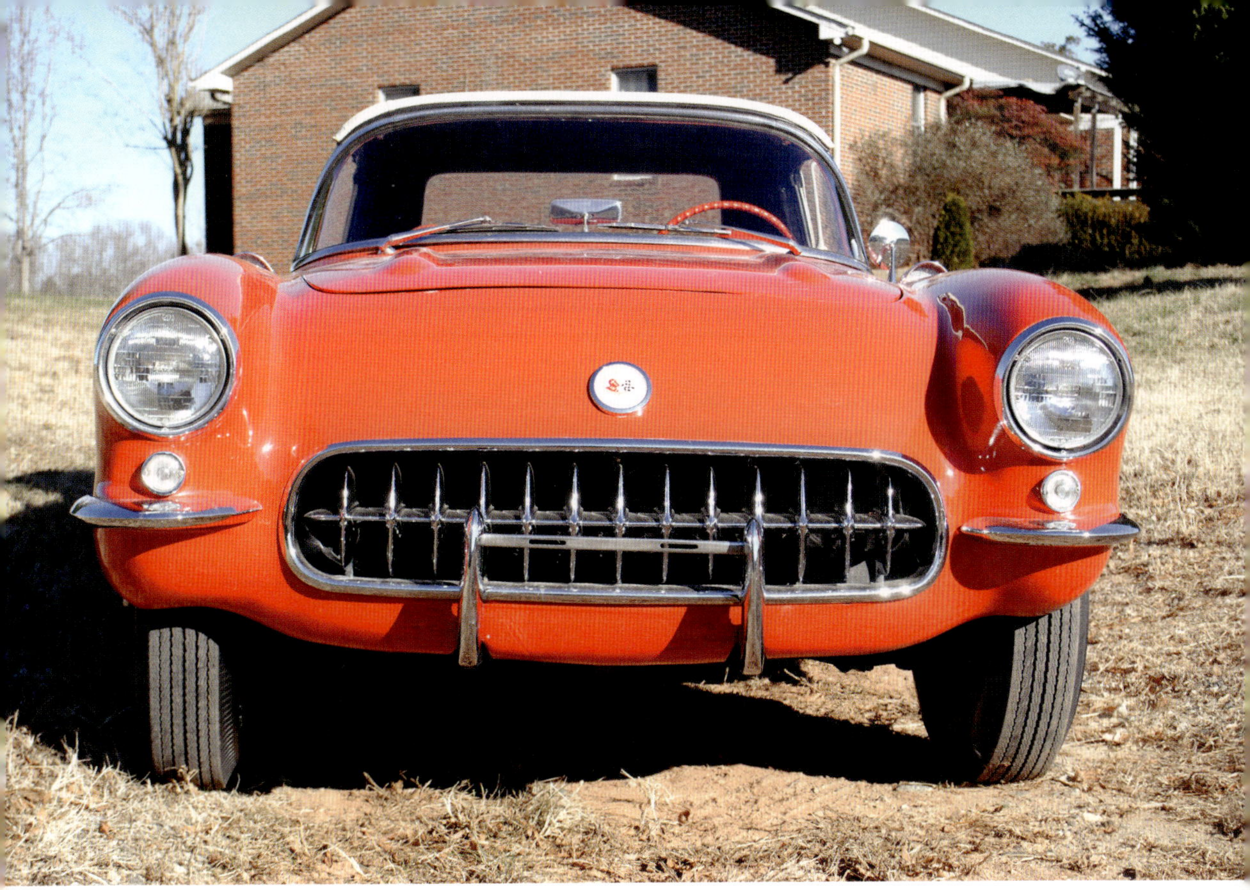

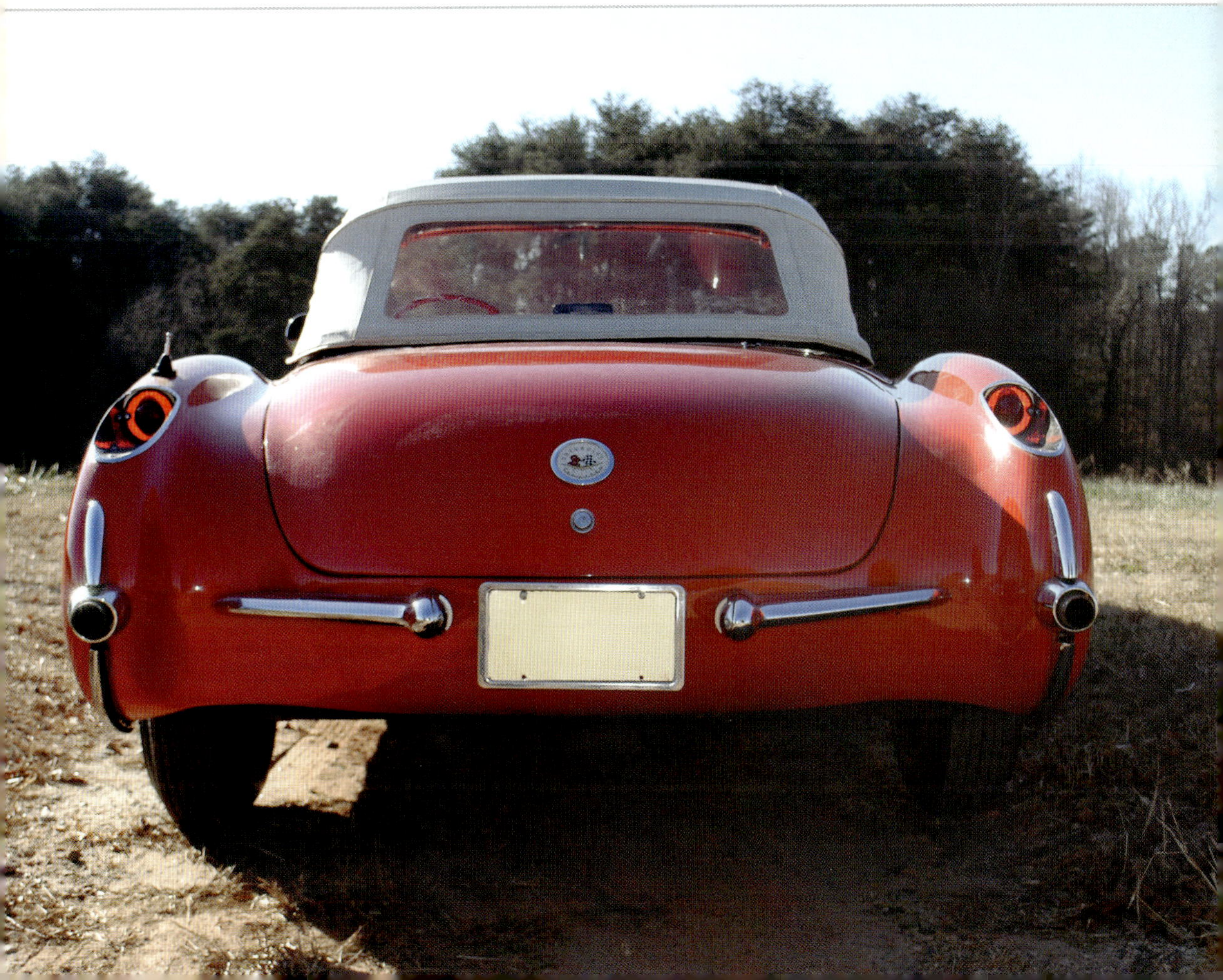

The Ultimate Ride

2015 Z06, Laguna Blue

Owners: Stan and Mary

Stan and Mary are the proud owners of this 2015 Z06 Laguna Blue Corvette. The Z06 with the Z07 package is the ultimate Corvette. This is their third Corvette. The first was a 1976 and the second was a 2008 Velocity Yellow Z06. Mary was all aboard for the 2015 Z06; she even traded in her 2008 car for the ultimate ride.

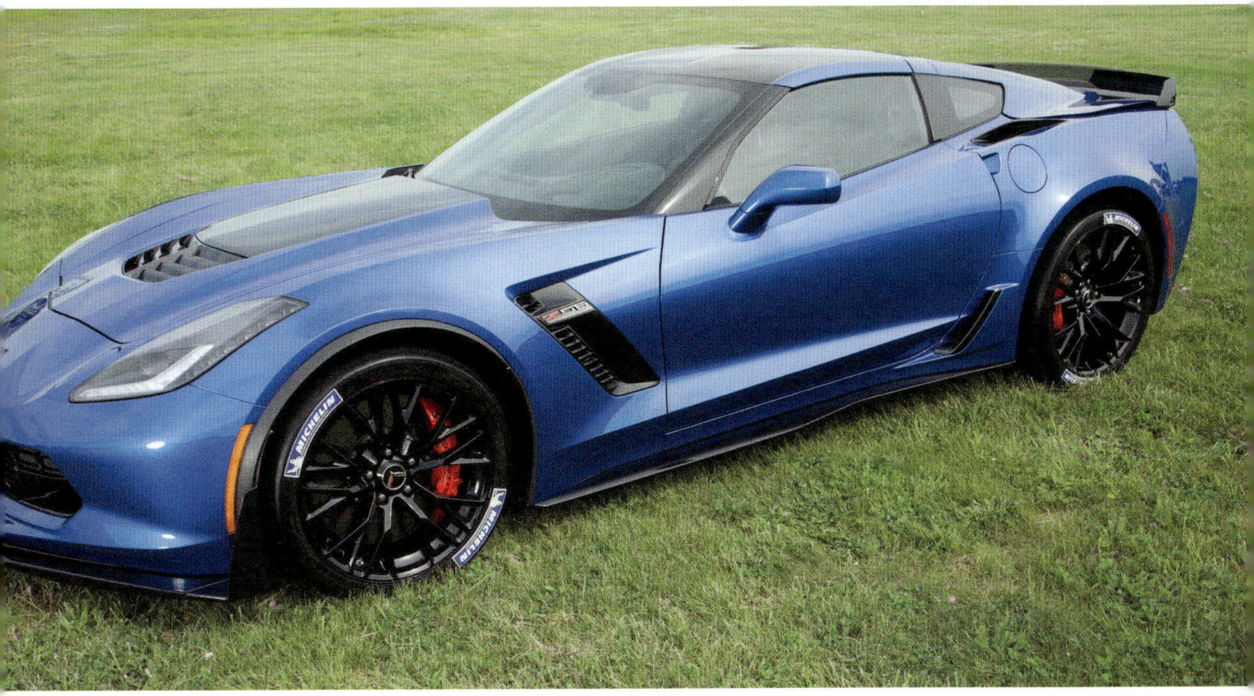
MICHELIN

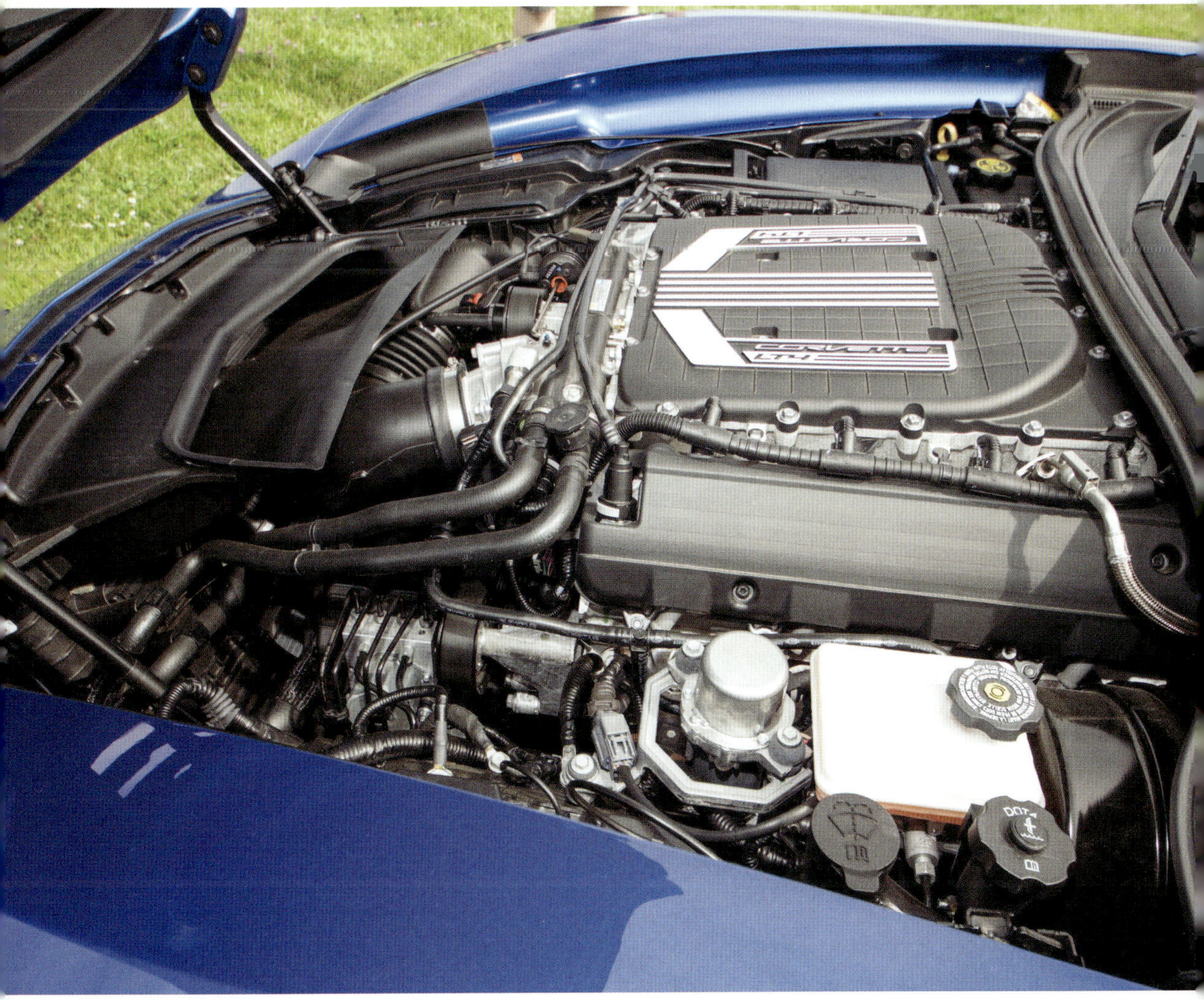

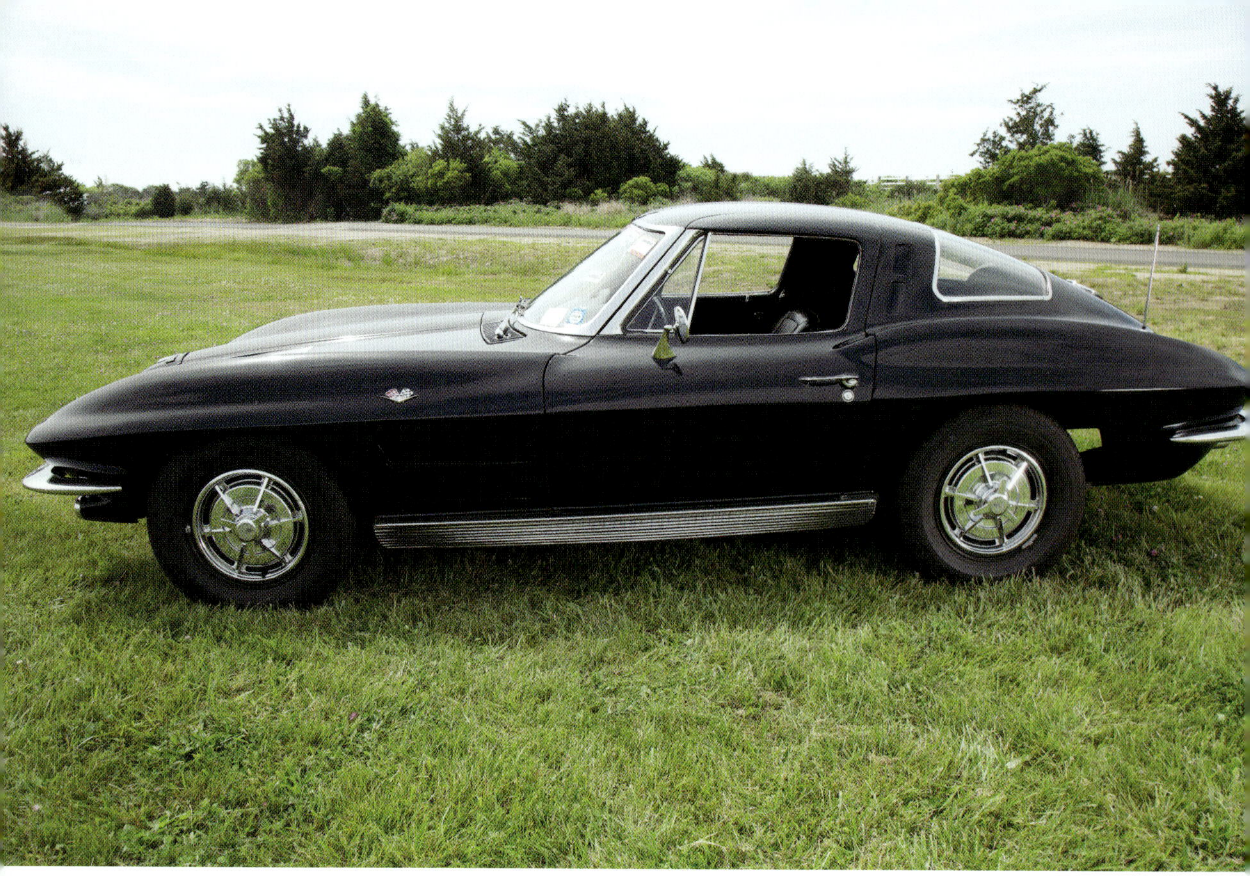

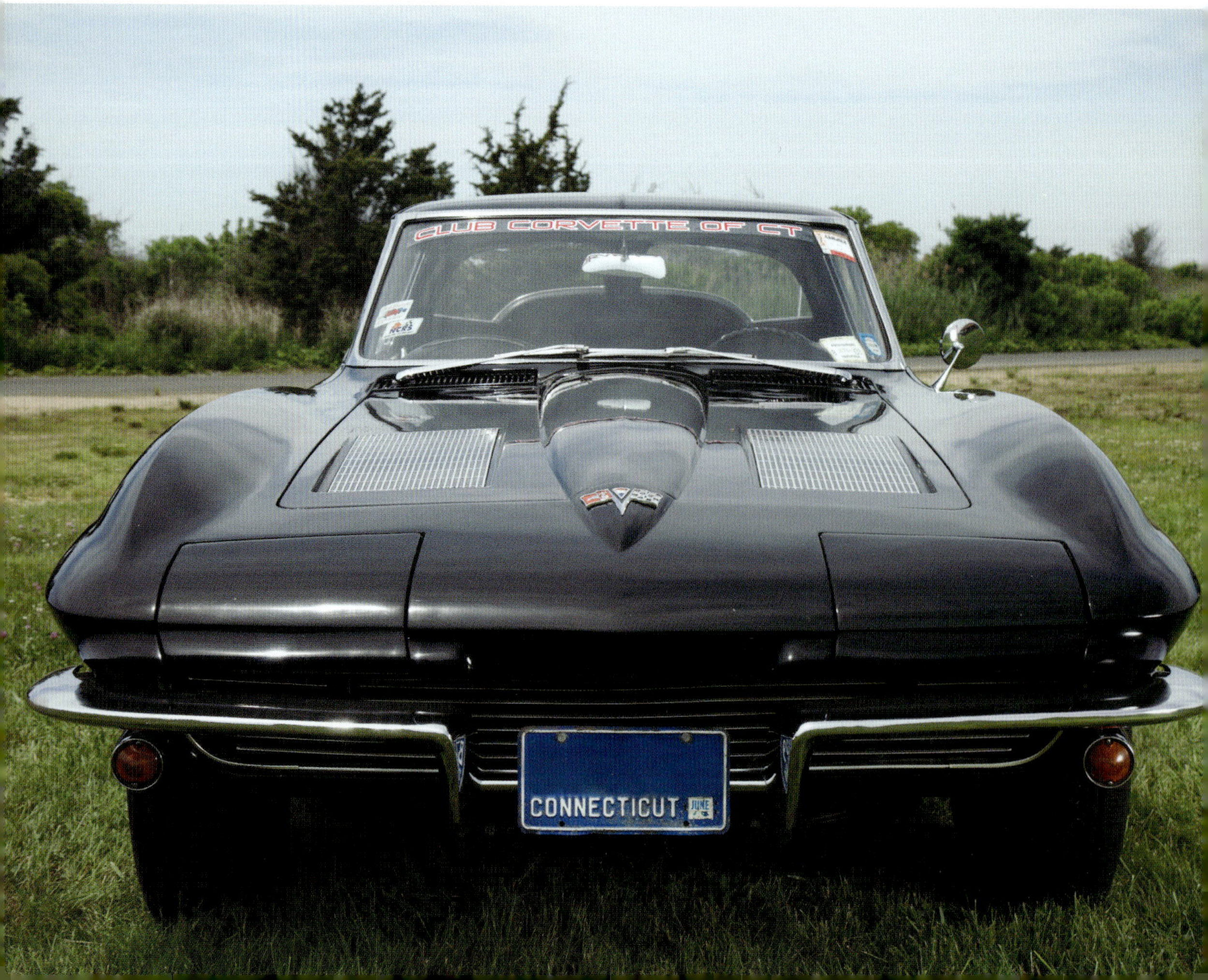
CLUB CORVETTE OF CT
CONNECTICUT

Nicknamed "Baby"

1963, Split Window, Black
Owner: Ivan

Ivan affectionately nicknamed his 1963 black split-window Corvette "Baby." He had been on the lookout for a black 1963 Corvette for 35 years because he had an identical one when he was 18, a gift from his grandmother. However, after he got married, his wife was unable to drive it so they made the decision to sell the car in 1967.

"I had always dreamed of buying another one, and in 2001, my cousin found Baby in upstate New York. Over the past several years, Baby has won best-in-class show trophies and was included in Corvettes at Carlisle, the 50th anniversary of the '63, and has been judged by NCRS," says Ivan. "Today it sits on its lift in my garage as my prized trophy."

Three Decades of Driving

1982, Silver Beige

Owner: George

The 1982 Silver Beige Corvette shown here is the first Corvette George has owned. In 1988, after college tuition payments were a thing of the past, he decided to purchase this beautiful machine on wheels. He specifically wanted a 1982 Collector Edition (C.E.), not for any special reason—he just did. He has driven it for almost 30 years and has never regretted one minute of ownership.

The 1982 was the last of the C3 generation (1968-1982) of Corvettes. The C.E. was a unique model to commemorate the end of the C3 generation and it incorporated special features not available on a standard 1982. The C.E. rear window is a "hatch," which opens to provide access to the trunk space. This model also had a special print and graphic combination with matching interior Silver Beige leather and fabric. The wheels are similar to the rib design of the 1967. The engine is a standard 350ci displacement, except the normal Corvette carburetor was replaced with a throttle-body fuel injection system. A total of 6,759 Collector Editions were produced and this was the first Corvette to break the $20,000 price barrier at $22,538.

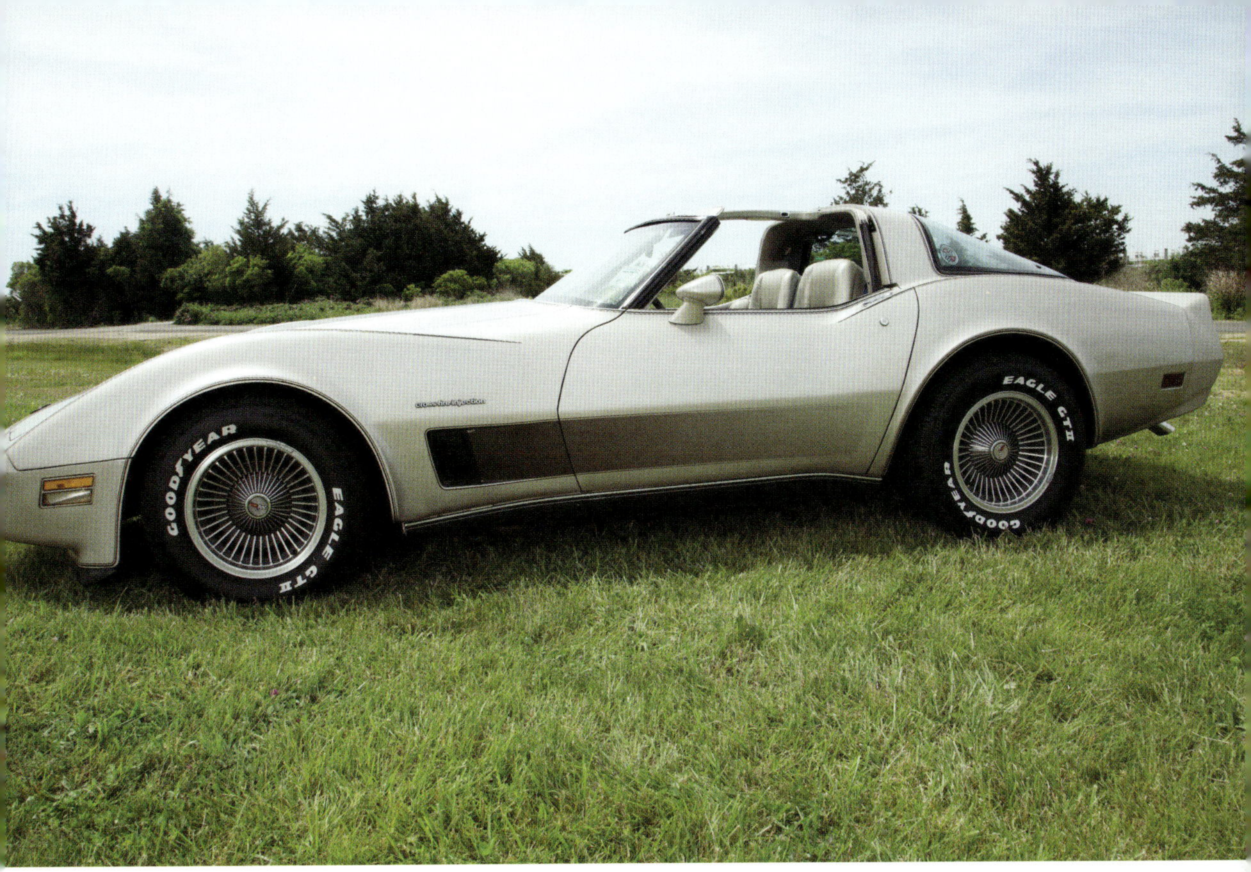
GOODYEAR
EAGLE GT II

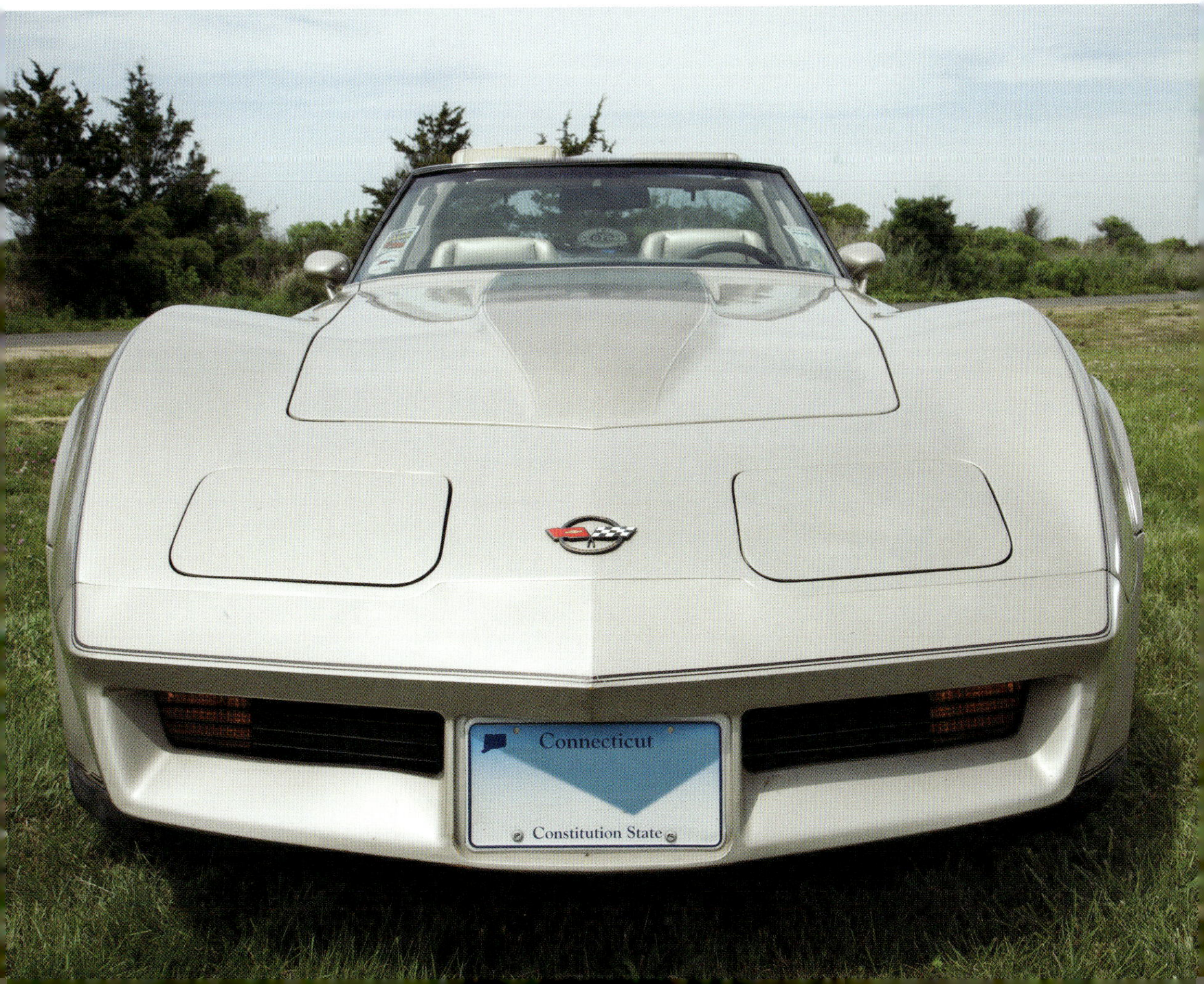
Connecticut
Constitution State

A $500,000 Corvette

1968 L-88 Coupe, LeMans Blue

Owner: Larry

Larry's 1968 blue Corvette is a factory L-88 coupe and one of only eighty made that year. The original cost of the car was $6,000; it is now valued at $500,000. The original L-88 included a special aluminum head and a compression race motor. This car is for racing and not for street use, which is why the radio and the fan shroud were deleted from the package.

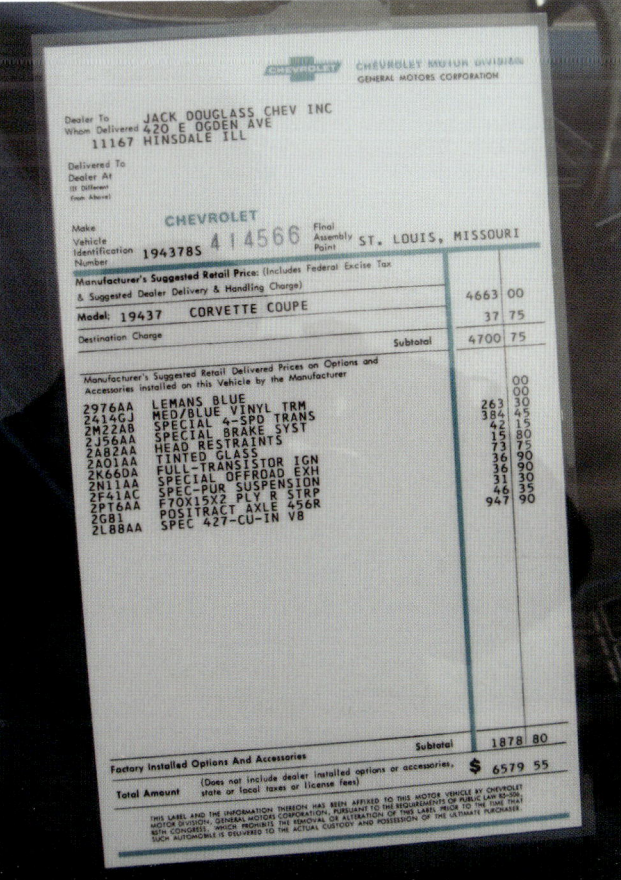

CHEVROLET MOTOR DIVISION
GENERAL MOTORS CORPORATION

Dealer To Whom Delivered   JACK DOUGLASS CHEV INC
                             420 E OGDEN AVE
                             11167 HINSDALE ILL

Delivered To Dealer At (If Different From Above)

Make   CHEVROLET
Vehicle Identification Number   194378S 414566   Final Assembly Point   ST. LOUIS, MISSOURI

Manufacturer's Suggested Retail Price: (Includes Federal Excise Tax & Suggested Dealer Delivery & Handling Charge)		4663	00
Model: 19437 CORVETTE COUPE		37	75
Destination Charge			
	Subtotal	4700	75

Manufacturer's Suggested Retail Delivered Prices on Options and Accessories Installed on this Vehicle by the Manufacturer

2976AA	LEMANS BLUE		00
			00
			30
24146CJ	MED/BLUE VINYL TRM	263	10
2M22AB	SPECIAL 4-SPD TRANS	384	45
2J56AA	SPECIAL BRAKE SYST	42	15
2A82AA	HEAD RESTRAINTS	15	80
2A01AA	TINTED GLASS	73	75
2K66DA	FULL-TRANSISTOR IGN	36	90
2N11AA	SPECIAL OFFROAD EXH	36	90
2F41AC	SPEC-PUR SUSPENSION	31	30
2PT6AA	F70X15X2 PLY R STRP	46	35
2G81	POSITRACT AXLE 456R	947	90
2L88AA	SPEC 427-CU-IN V8		

	Subtotal	1878	80
Factory Installed Options And Accessories			
Total Amount (Does not include dealer installed options or accessories, state or local taxes or license fees)	$	6579	55

THIS LABEL AND THE INFORMATION THEREON HAS BEEN AFFIXED TO THIS MOTOR VEHICLE BY CHEVROLET MOTOR DIVISION, GENERAL MOTORS CORPORATION, PURSUANT TO THE REQUIREMENTS OF PUBLIC LAW 85-506. WHICH PROHIBITS THE REMOVAL OR ALTERATION OF THIS LABEL PRIOR TO THE TIME THAT WITH CONGRESS, SUCH AUTOMOBILE IS DELIVERED TO THE ACTUAL CUSTODY AND POSSESSION OF THE ULTIMATE PURCHASER

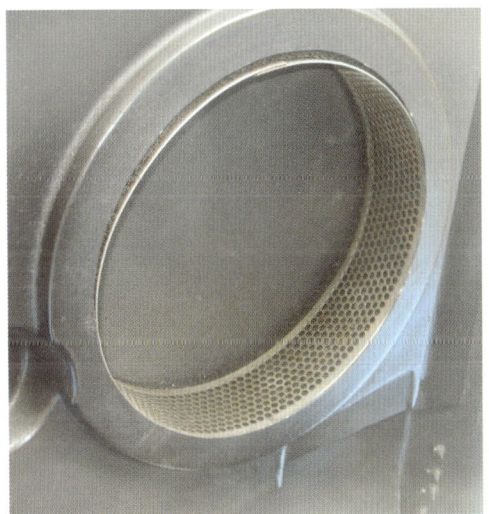

A Frame-Off Restoration
1959 Convertible, Inca Silver with White Cove
Owner: Patrick

Over the years, Patrick has owned a number of Corvettes. His first was a 1963 Daytona Blue convertible purchased in 1967; that was when he became addicted to Corvettes. In 2013, he purchased this 1959 Inca Silver with a white cove and red interior as a project car from the original owner. It had been stored in a barn in Missouri since 1975. Once the car arrived in Connecticut, Patrick and his wife realized this was a rare fuel-injected model, with both hard and soft tops. It needed to be preserved and restored to originality according to NCRS (National Corvette Restorer Society) standards.

They spent two years working out of their garage, completing a frame-off restoration, which included rebuilding the engine, replacing every nut and bolt that was not original to the car, carpeting, upholstery, convertible, and hard tops, stripping, painting, and more!

In 2015, they received a Top Flight award from their local chapter; in 2016, they attended the NCRS National Convention in Rhode Island where they earned the National Top Flight Award. "We have enjoyed showing this iconic 1959 Corvette in local shows and have been very proud to come away with awards every time," says Patrick.

First Corvette

2013 Z06, Artic White

Owner: Sandy

Why a Corvette? Sandy explains, "My introduction to Corvettes was when I was seven years old in 1963. My father took me to the GM dealership in Cleveland, Ohio. It was one of the rare times I was allowed to roam around by myself. I walked into the Chevy area and saw a blue Corvette split-window coupe. I ran up to it and immediately mushed my face and hands against the side window. The salesman asked if I would like to sit behind the wheel. When I did, I felt like I was in Buck Rogers' spaceship. It took me 51 years before I could afford one I really wanted. This 2013 Z06 is my first Corvette and it is unbelievable to drive, *especially* with Michelin Pilot Super Sport tires. That was the best $1700 upgrade I could have made. This car is comfortable *and* practical enough to drive cross-country—or

I can take it to a race track where it will run away from just about anything that is not heavily modified for racing."

Sandy's car is one of only 128 Arctic White Z06s, and one of only 64 with the base 1LZ interior option. The total number of of C6 Z06s made was just under 28,000.

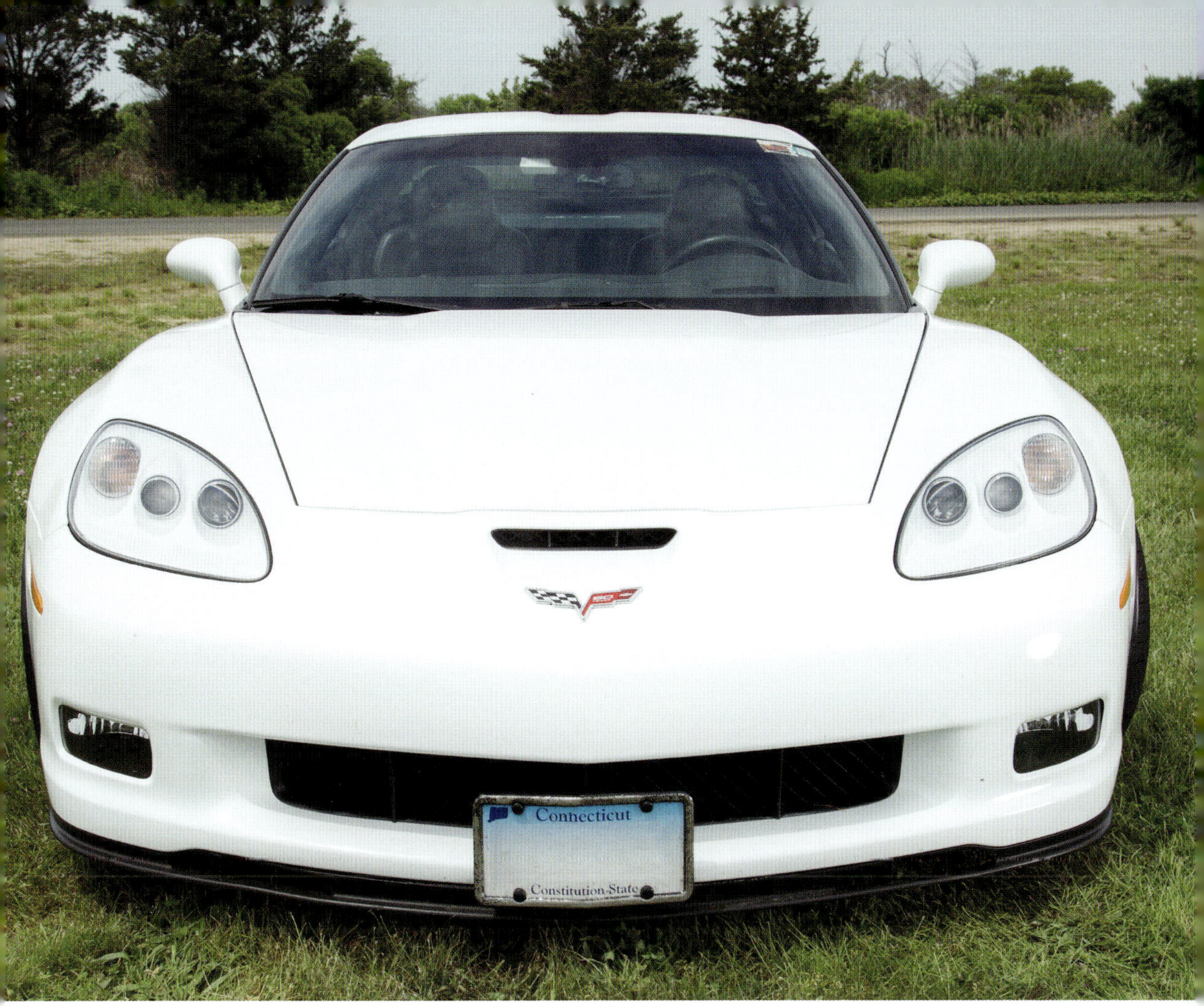
Connecticut
Constitution State

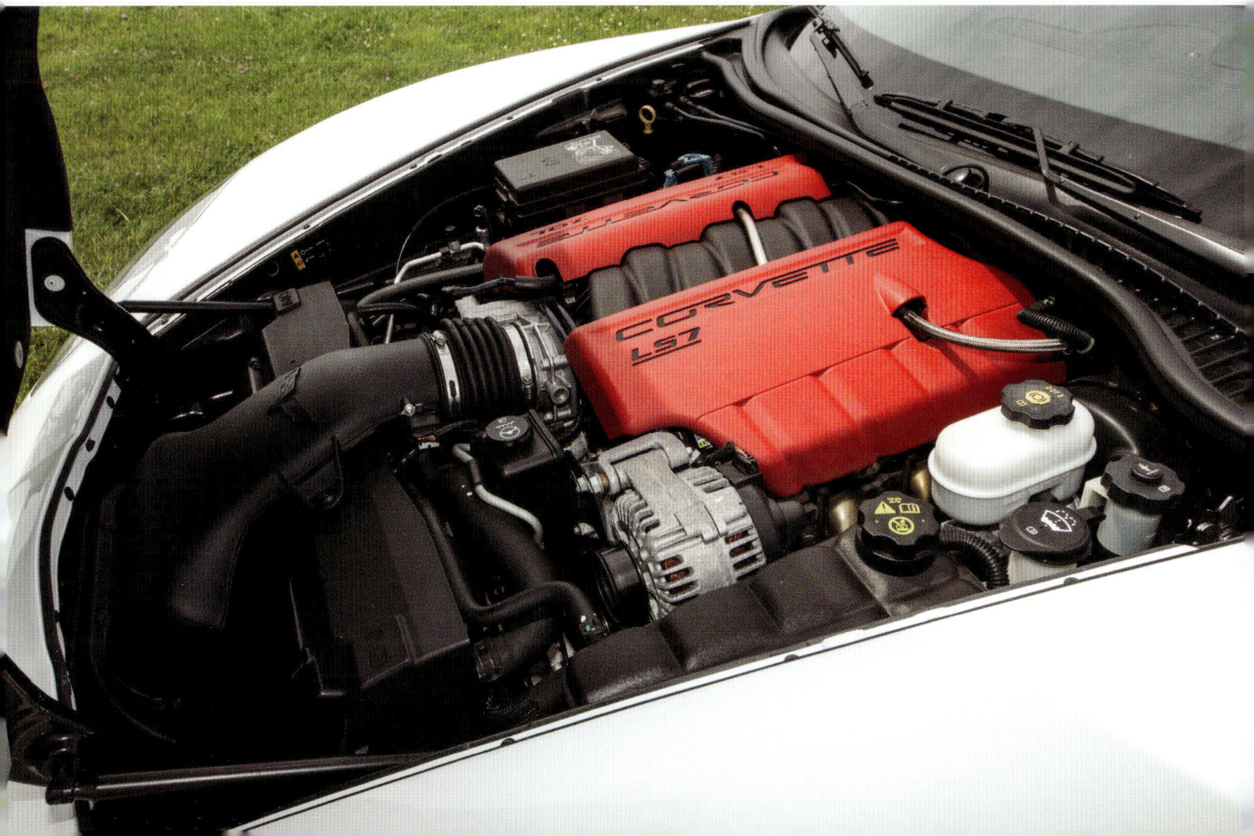
CORVETTE
LS7

A Rare Combination

2016 Stingray Z51, Night Race Blue

Owner: George

The rare combination of the exterior color (Night Race Blue) with the convertible top color (Night Shadow Blue) and the interior (Brownstone leather) applied to a Stingray Z51 convertible made this Corvette unique for the model year 2016.

This special Corvette (born on June 30, 2015, in Bowling Green, Kentucky) was christened MARCO by his owner, George, in honor of the famous explorer. True to his name, before his second birthday MARCO had traveled across North America, and logged nearly 50,000 miles. Like his older brother before him, WALDO (a 2004 Magnetic Red convertible approaching 300,000 miles), MARCO will seek to visit all 49 continental states and each of the Canadian provinces.

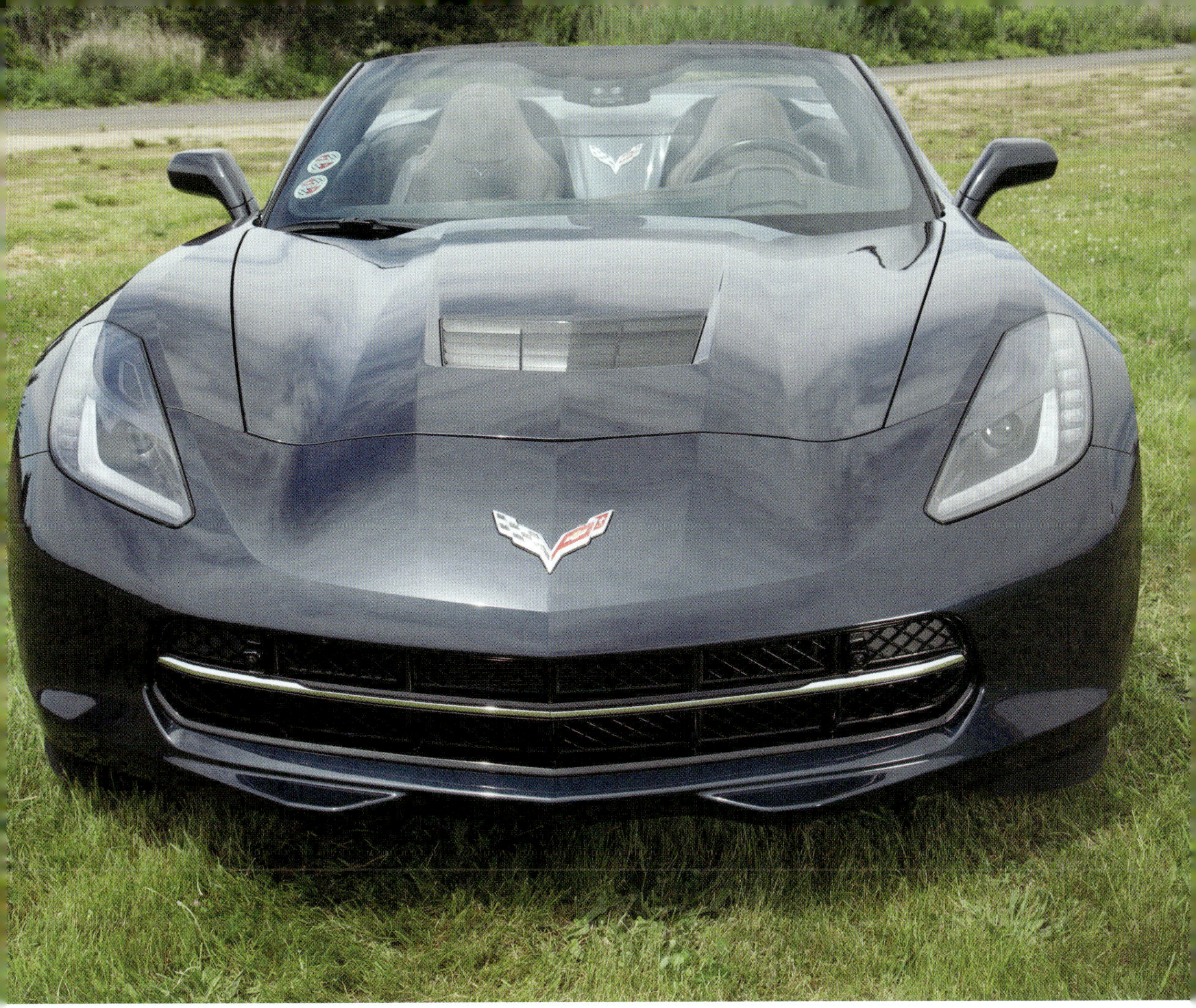

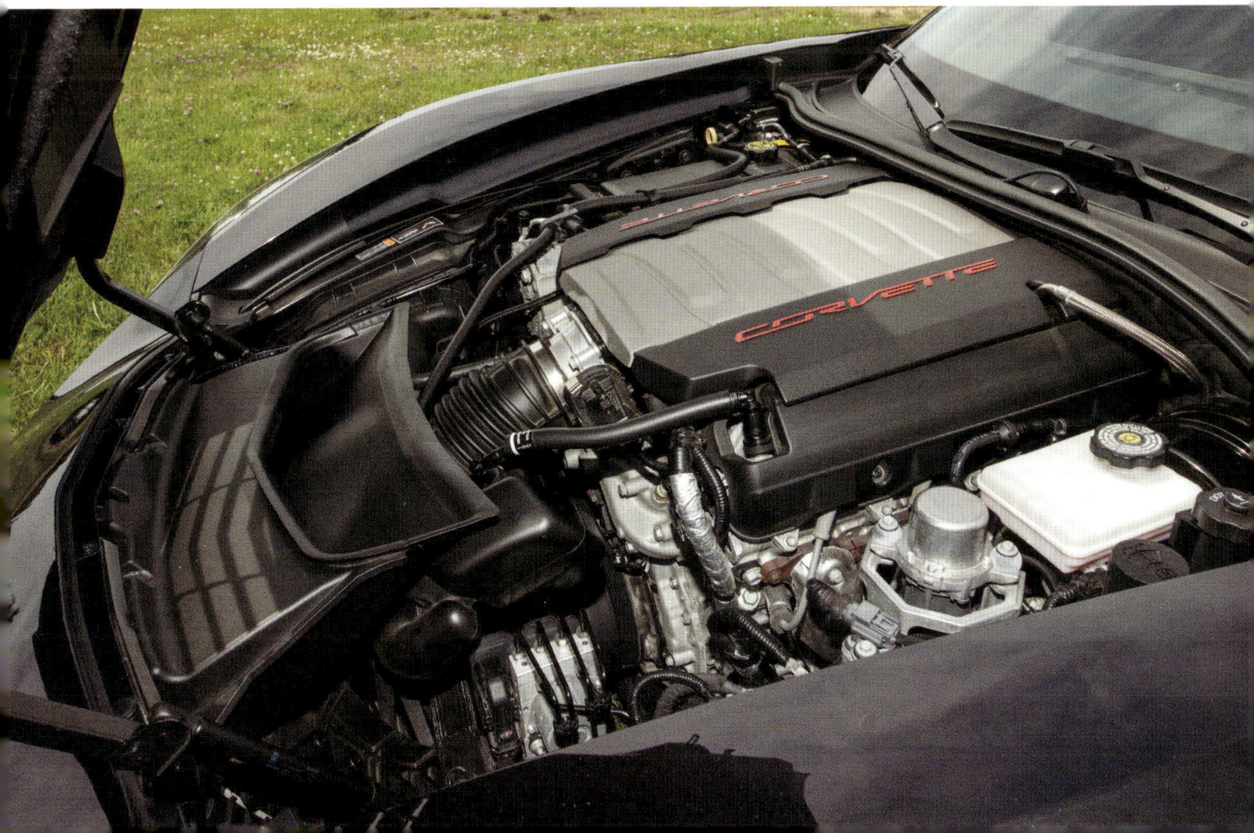
CORVETTE

One Owner for Fifty Years

1967, Ermine White

Owner: Jim

Jim and his wife purchased this Corvette brand new in 1967 while living in New Jersey. In the photos below, his son is behind the wheel in 1967 and fifty years later in 2017.

Asked about the car, Jim says, "While living in Rochester, New York, in 1969, I put studded tires on all four wheels and drove the Corvette through some of the worst ice and snow storms in Rochester's history. In 1971 we moved to Springfield, Massachusetts, and the Corvette was stored in my parent's garage for the next eighteen years!

"In 1989, I sent it to a body shop where most of the rubber (trim and molding) was replaced with OEM parts. The body was repainted with the factory Ermine White formulation. The interior is all original and the odometer currently reads 89,000 miles. The Corvette is stored six months out of the year in a heated garage."

The Perfect Color

2002 C5, Electron Blue
Owner: Jerry

Jerry searched for a C5 Corvette, which was manufactured from 1997 to 2004; this was the last era to have hidden headlights. When he saw the color, Electron Blue, which was only made in 2002 and 2003, he knew this was the color for him.

After much searching, Jerry finally found a 2002 Electron Blue, which still remains his favorite Corvette color. It's a fun car to drive, and truly a "garage queen."

"Whenever we are on the road and see another Corvette, we always wave to each other; this is called 'save the wave'," he says.

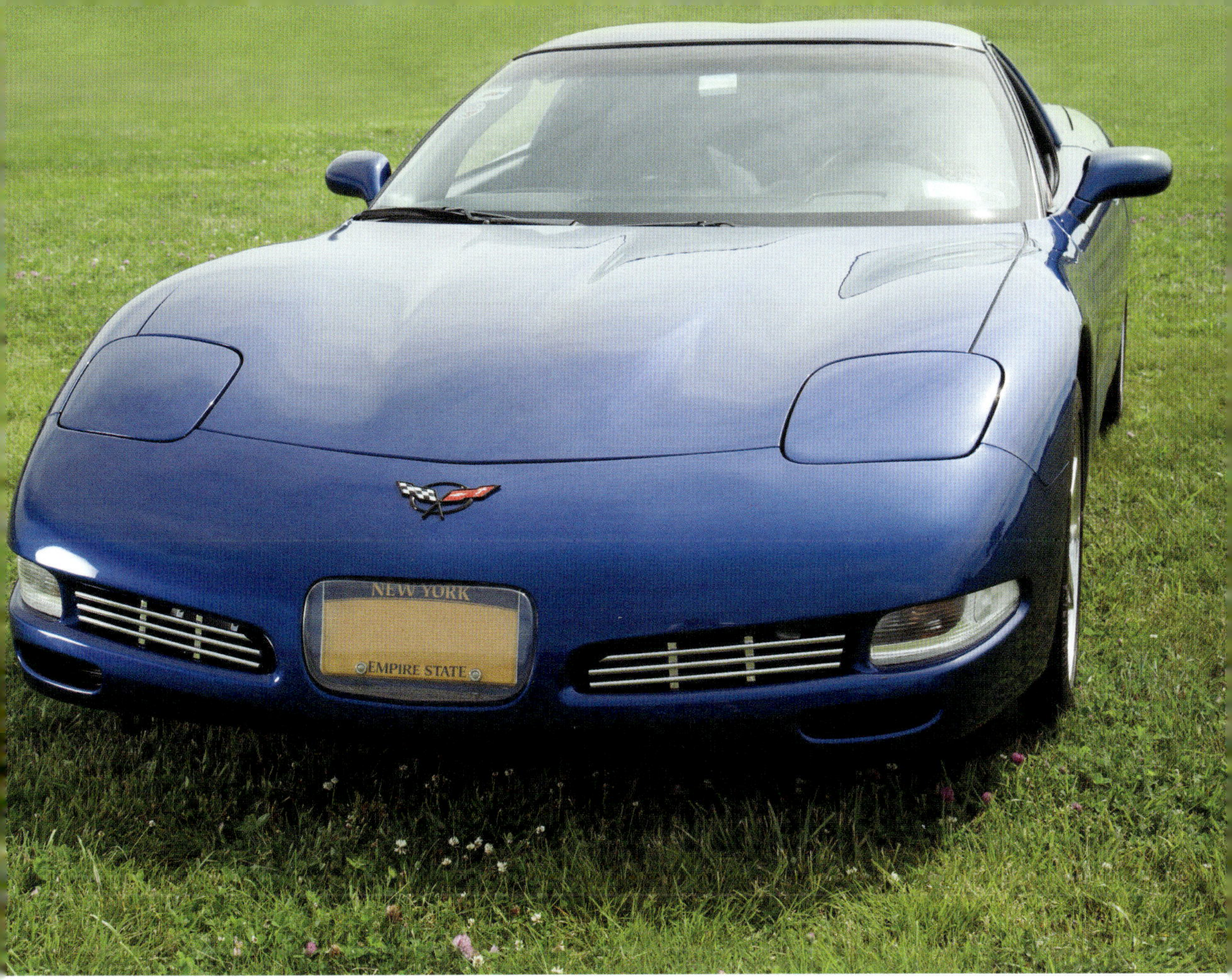
NEW YORK
EMPIRE STATE

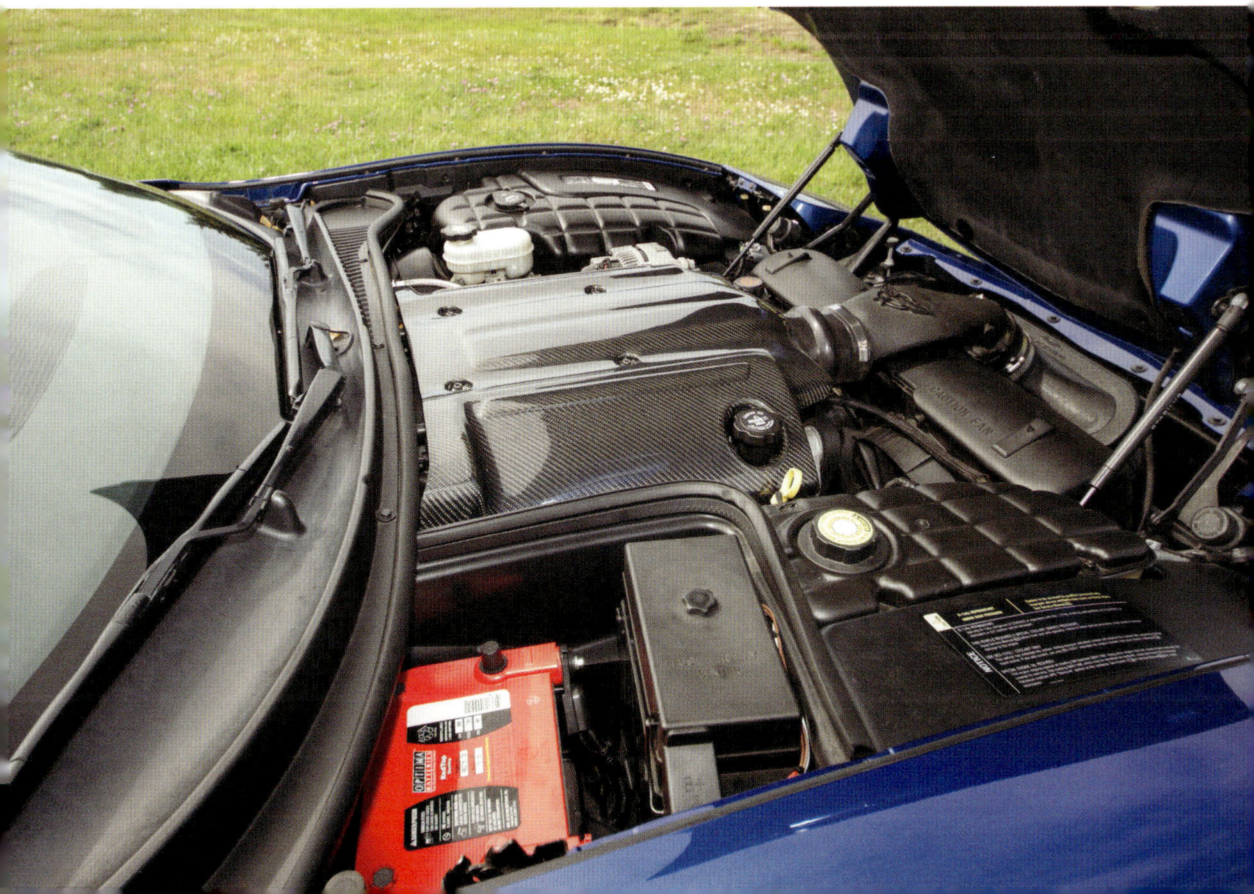

![Black 2008 Corvette convertible parked on grass]

Join the Club

2008 Convertible, Black

Owner: Patricia

This 2008 convertible is Patricia's first Corvette, but her husband Michael once owned a 1961 model. They love their fun car and enjoy belonging to their state Corvette Club because of the nice people and the good times at car shows and picnics.

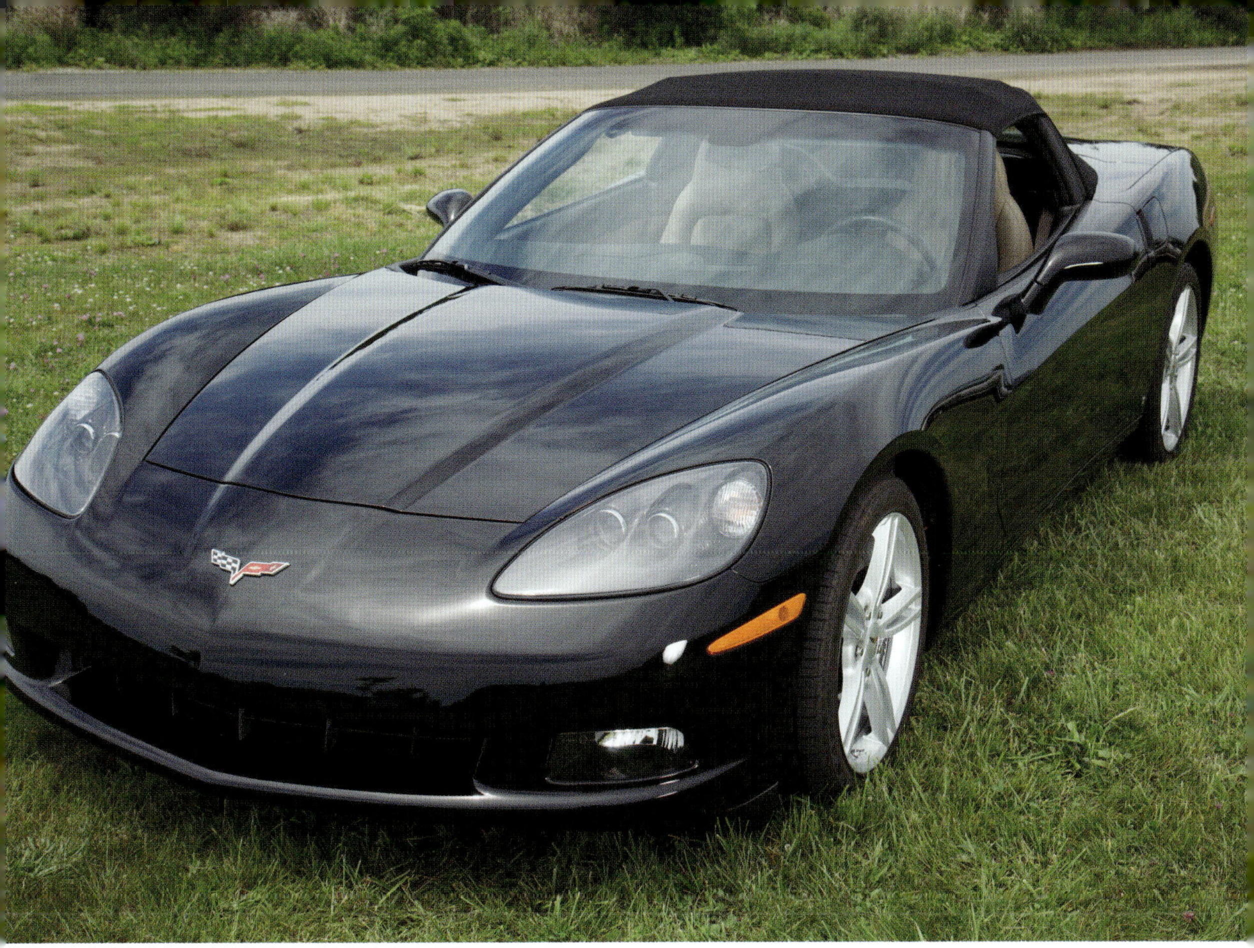

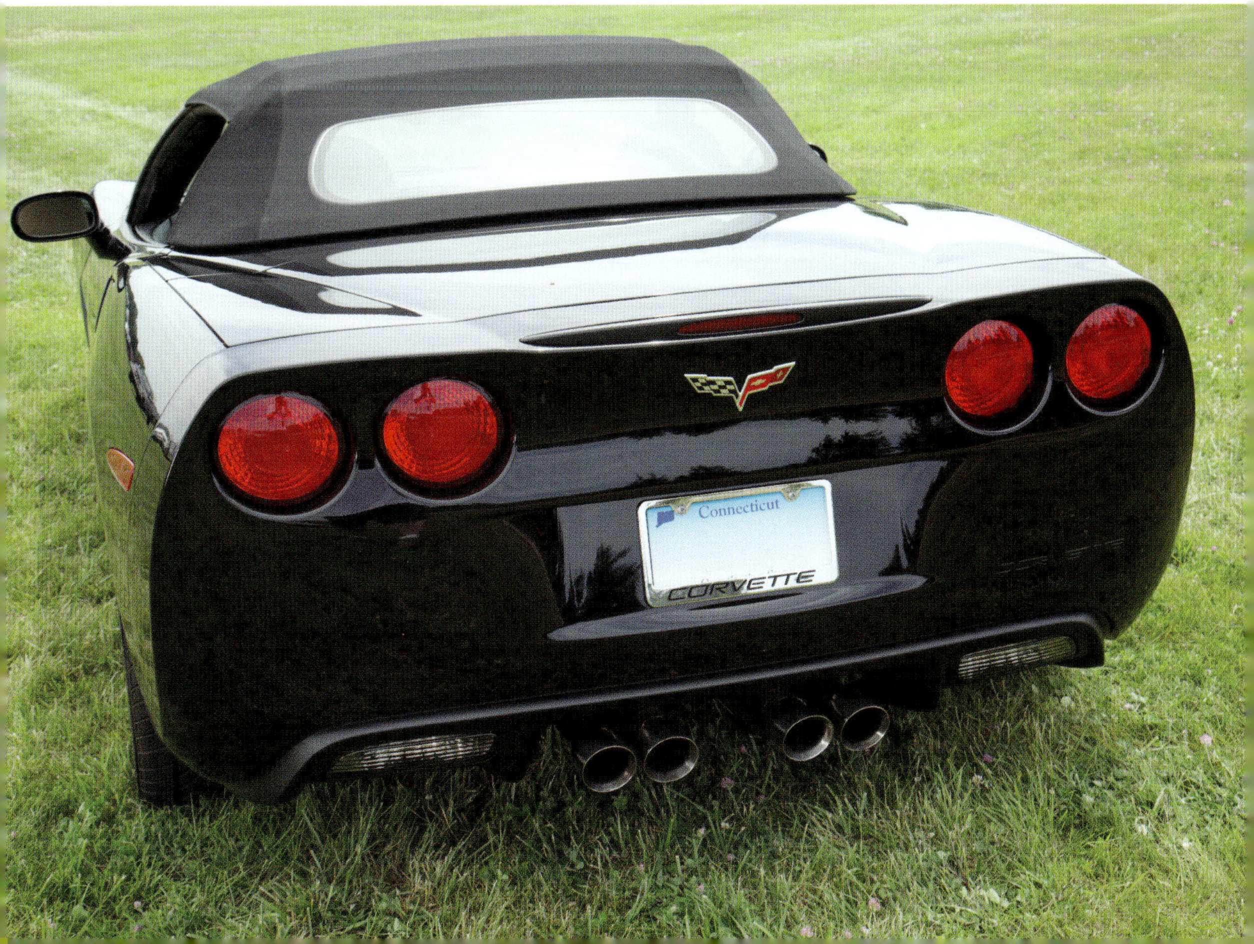
Connecticut
CORVETTE

In Touch with History

1966 Convertible, Nassau Blue/Bright Blue
Owner: Dave

Dave's first exposure to a Corvette was in 1973; he was thirteen years old and a friend's older brother gave him a ride in his 1969 Riverside Gold Stingray. Dave bought his first Corvette eight years later, a 1980 L-82. He kept the car for nine years, then sold it for a new 1990 coupe, which he held onto for twenty years. In 2010, he took the plunge to buy a long desired midyear Corvette. He selected this Nassau Blue/Bright Blue 1966 convertible with assistance from a NCRS judge.

"I tracked down the original salesperson and the original owner of the car," says Dave. "The owner was happy to hear that the car was still running and he sent a picture of the car from 1966. He confirmed that he pampered the car for thirty years before selling it to the second owner. This car is a NCRS National Top Flight car and is a thrill to drive."

He adds, "I am now in my 37th year of Corvette ownership and love the feeling of Americana that comes with every Corvette, as well as the history of the vehicle. For me, the perfect garage would contain a midyear convertible, a 1963 split window coupe (with air conditioning), and a C7 daily driver. I still have two Corvettes to go!"

Seize the Moment

2006 Convertible, Velocity Yellow
Owner: Gary

Gary's passion for Corvettes runs deep. Despite many adversities over the years, he is excited that he was finally able to purchase this special car.

When he was five years old, Gary ate a bag full of medications. "I literally died, and had to be brought back." Later, when he and now-wife Melissa started dating, he was in a bad motorcycle accident; if he had not been wearing a helmet, he probably would not have survived. More recently, he was diagnosed with cancer. "I have learned to look at life from a different perspective. If you want something, get it! Enjoy it!" he says.

Gary went on to say, "This 2006 Velocity Yellow Corvette is a gift from me to me. I was going to back out of the sale to use the money for other things, but my wife grabbed me by the arm and said, 'Do it!'"

Gary adds, "I cherish life and I live it day-to-day, because you never know what may happen tomorrow. I am fortunate to have survived and to have my wonderful wife and my seven beautiful children. I cannot be happier and I never look back. Who knows, maybe one day I'll go for a Z06."

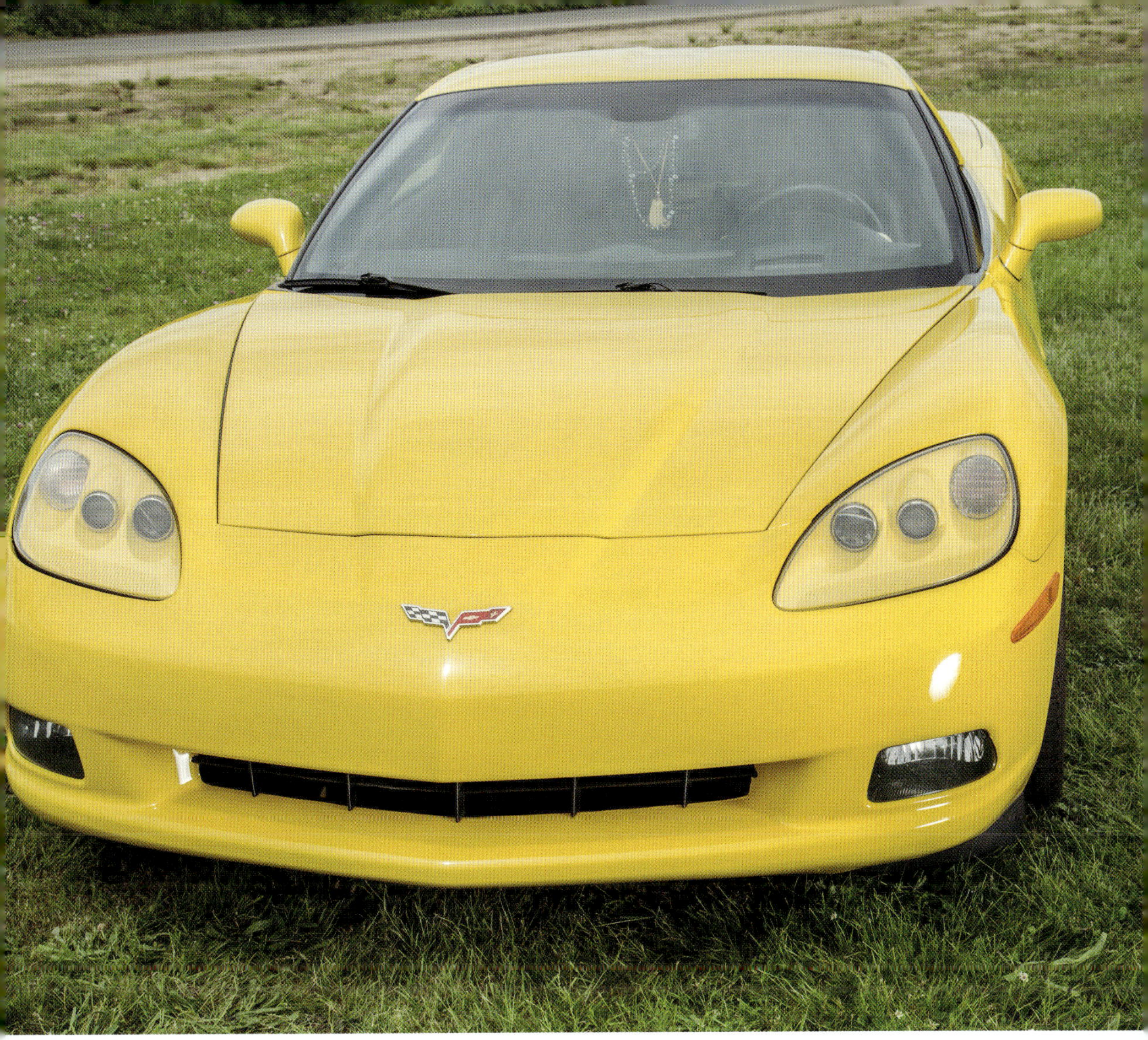

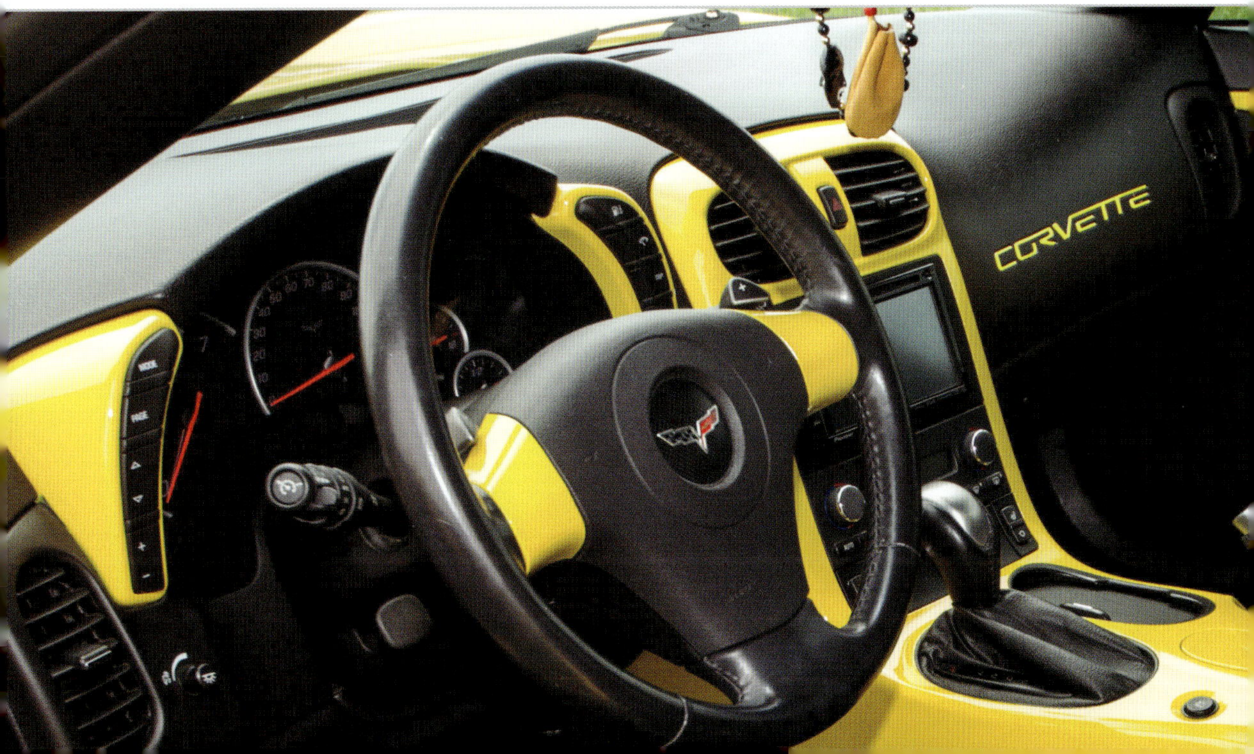
CORVETTE

A Collector

2016 Grand Sport, Supersonic Blue

Owner: Anthony

Anthony is a collector of Corvettes; this is one of three that he currently owns. This 2011 Corvette Grand Sport, in Supersonic Blue, came from Arizona. He purchased it while he was on vacation and had it shipped back to the east coast. His two other Corvettes are a 1991 ZR1 and a 1991 coupe, both Turquoise Metallic.

Anthony's first Corvette was a light blue 1970 that he purchased from a friend for $3,500.

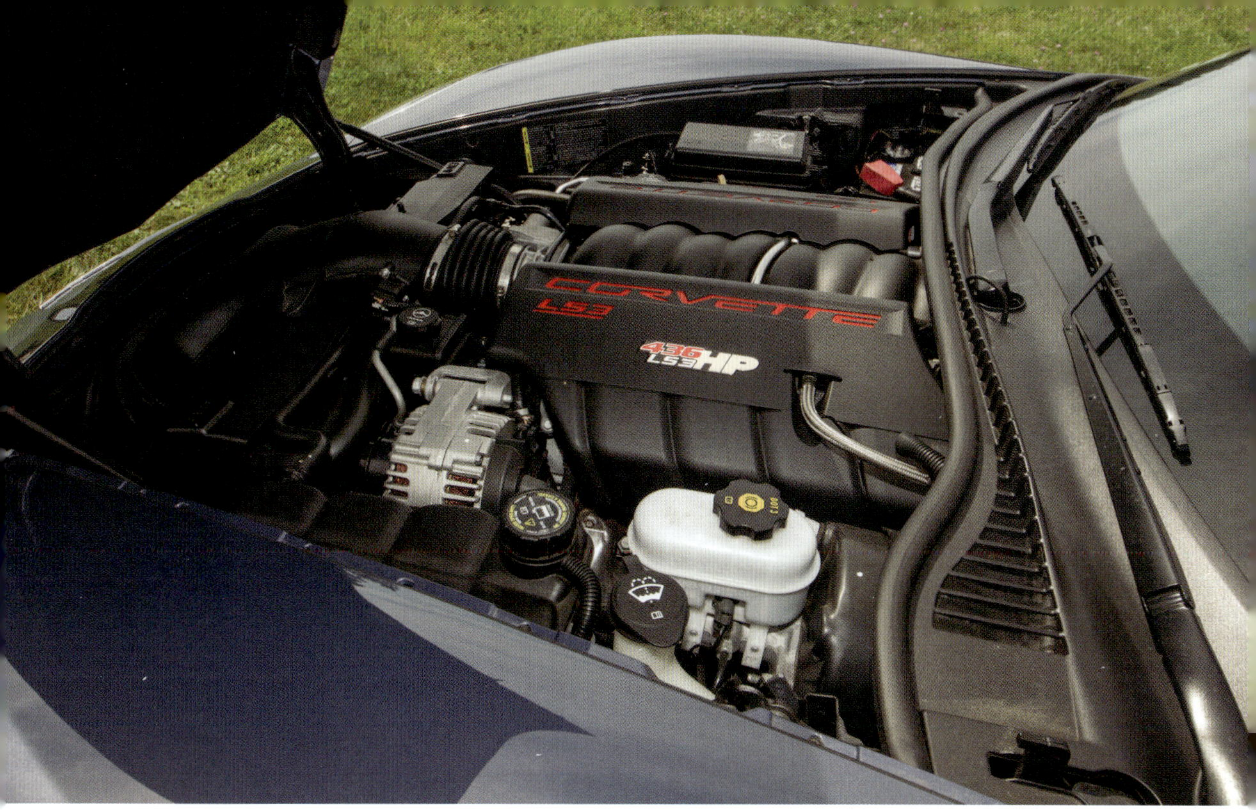
CORVETTE
LS3
436 LS3 HP

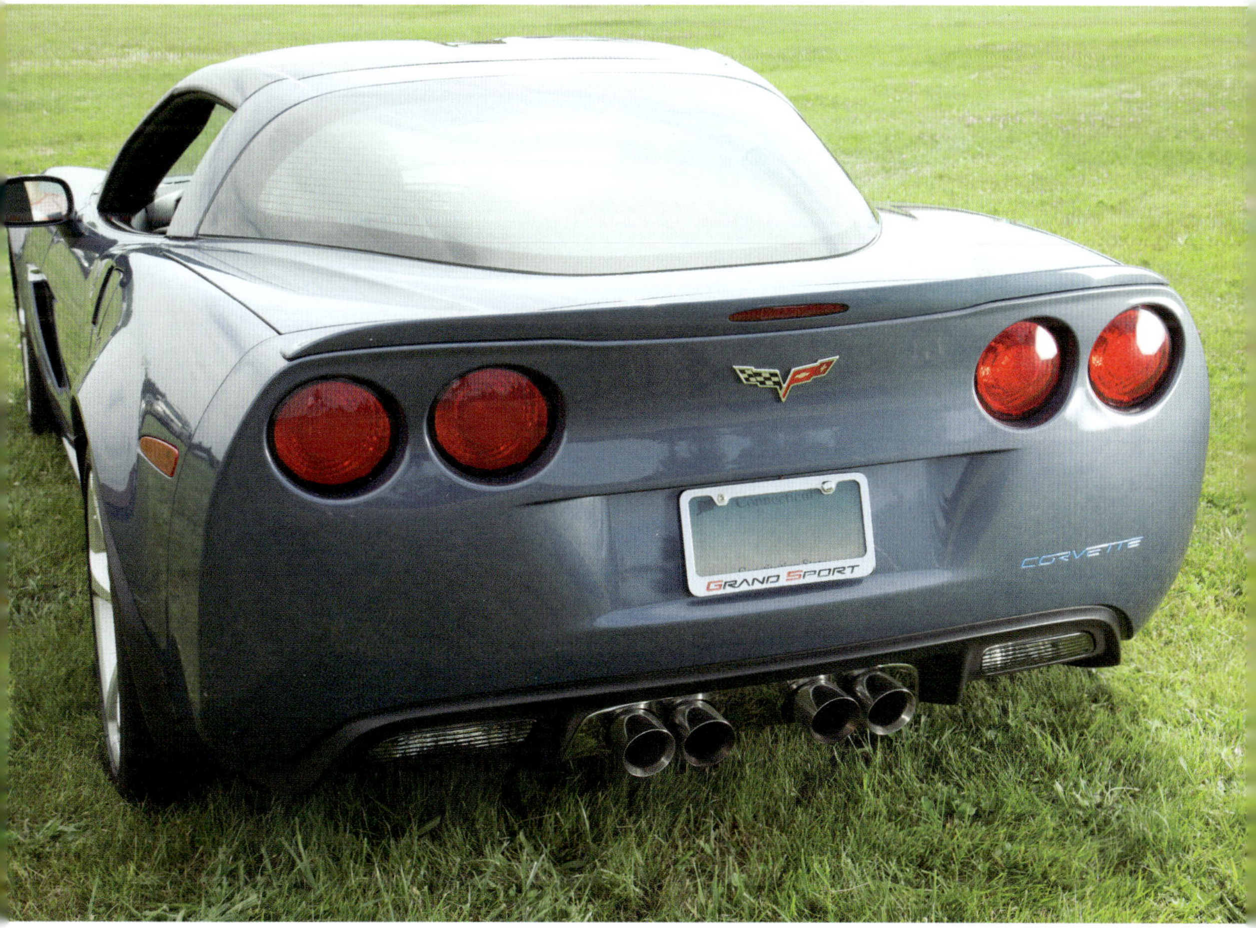
GRAND SPORT
CORVETTE

First Generation

1962 Restomod, Burgundy

Owner: Arthur

This 1962 Corvette is Arthur's first. After attending many car shows and cruise nights looking at the beautiful classic cars, he thought he wanted a 1955-1957 Thunderbird—but someone he met at a show convinced him to look at first generation Corvettes. "I saw a Corvette that had won the Cruiser of the Night Award and it had a small 'for sale' sign on it," he remembers. "It was a 1962 Restomod in pristine condition. The Restomod appealed to me because it had all of the features of a modern car with the look of the 1962. This was the best purchase I ever made. There is nothing like the camaraderie of classic car owners."

Route 66

2003 Convertible, Cyber Gray

Owner: Jack

Jack, like many other people, fell in love with Corvettes watching the television show *Route 66*, which aired from 1960–1964 and starred Martin Milner, George Maharis, and Glenn Corbett.

"This 2003 Cyber Gray Corvette is my fourth," says Jack. "My first was a 1968 Roadster that was waiting for me when I came home from Vietnam in March of 1969. I also owned a 1969 coupe followed by a 1988 coupe." Jack has been an active member of his state Corvette club since 2000 and is currently serving his second stint as president of that organization.

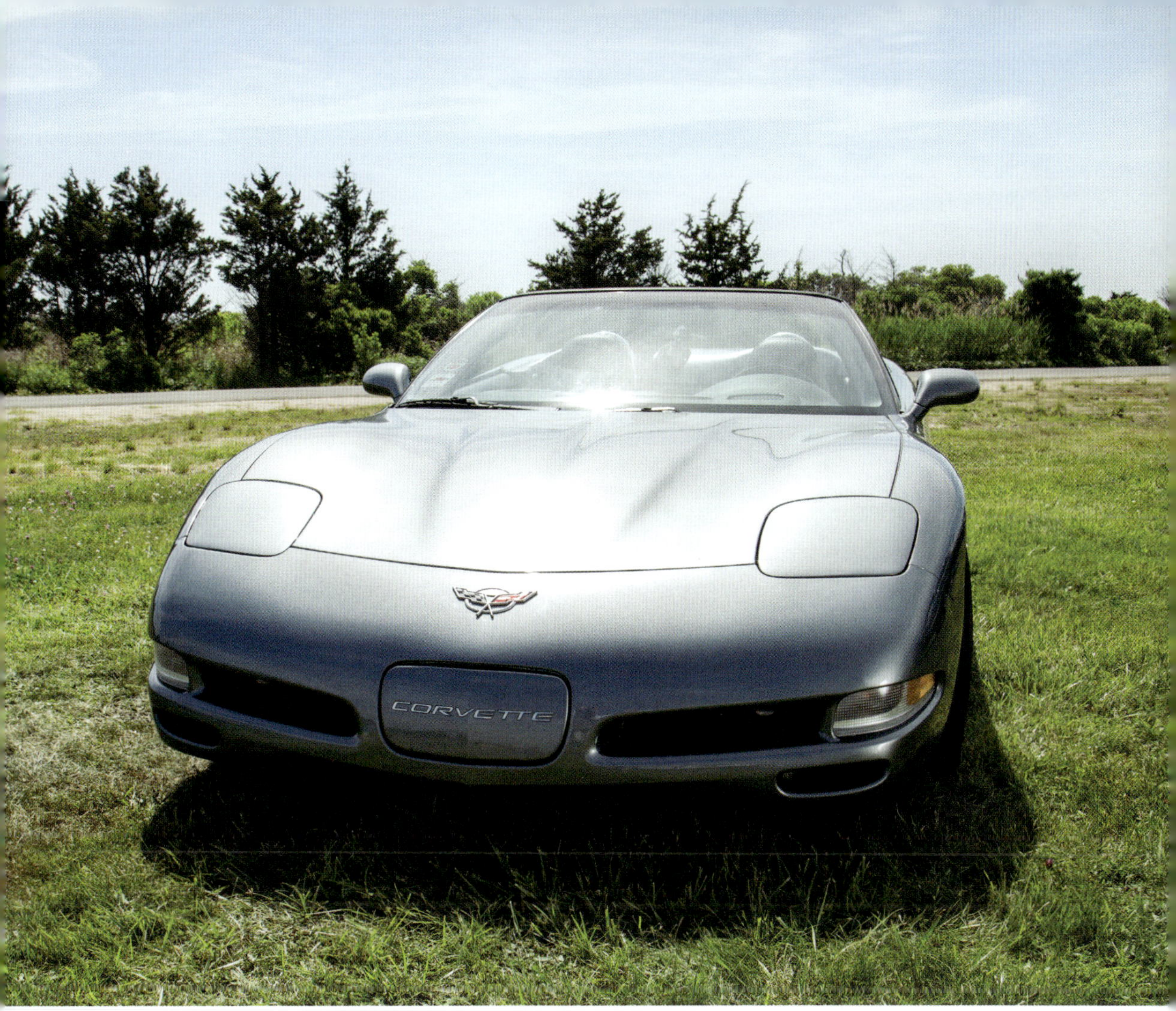
CORVETTE

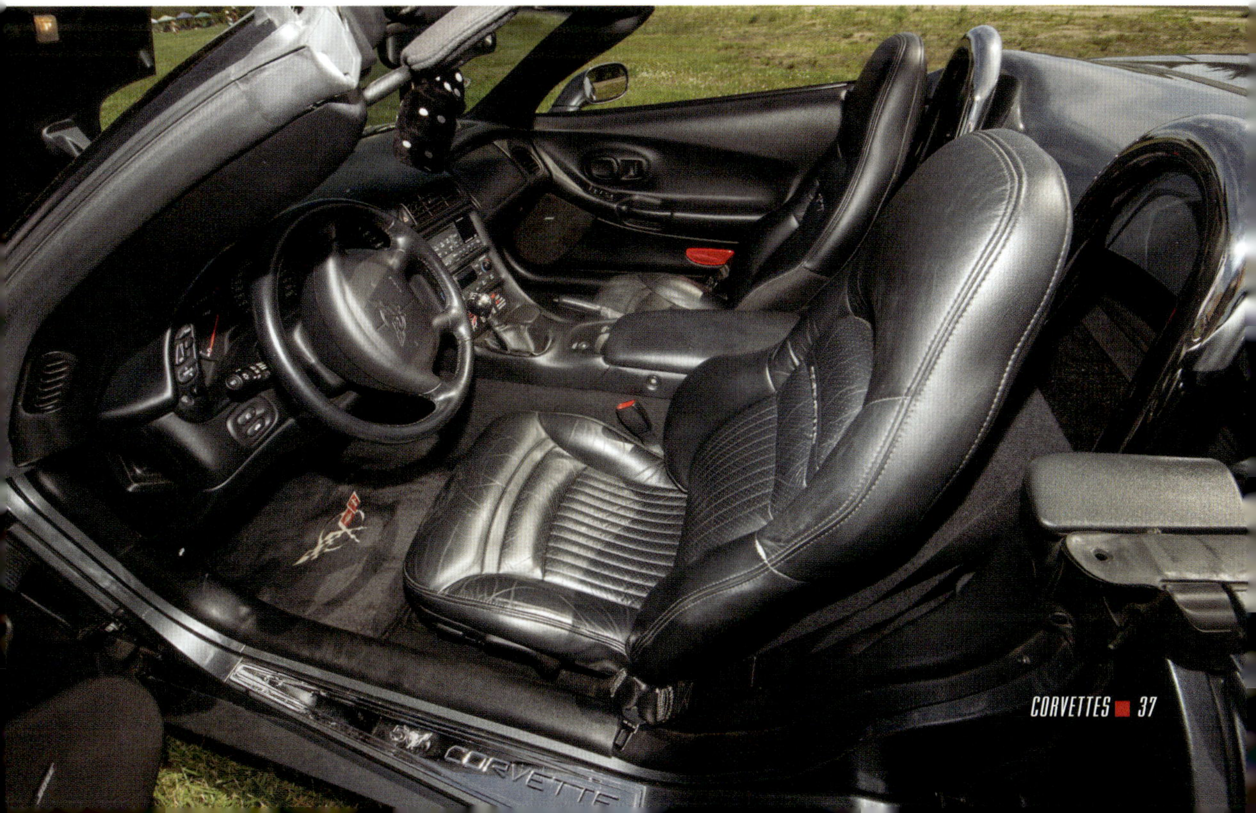

![Silver 1963 Corvette Split-Window Coupe parked on a leaf-covered lawn with trees in the background.]

This One Is Here to Stay

1963 Split-Window Coupe, Saddle Tan
Owner: Chuck

Chuck grew up in the early 1970s, a time when *any* Corvette on the road caught his attention. He always wanted to own one, beginning from a very early age. "When I was 28, I sold my motorcycle for a safer hobby and bought my first Corvette," he remembers. "It was a 1968 small block that needed a lot of work, but it was mine!"

Many years later buying and selling various Corvettes, Chuck found the epitome of all Corvettes: his 1963 Saddle Tan split-window coupe. This is possibly the most recognized Corvette in the history of these cars. "I don't ever say *never,* but I can't see myself selling this one anytime soon," Chuck says.

CHEVROLET
327
TURBO-FIRE

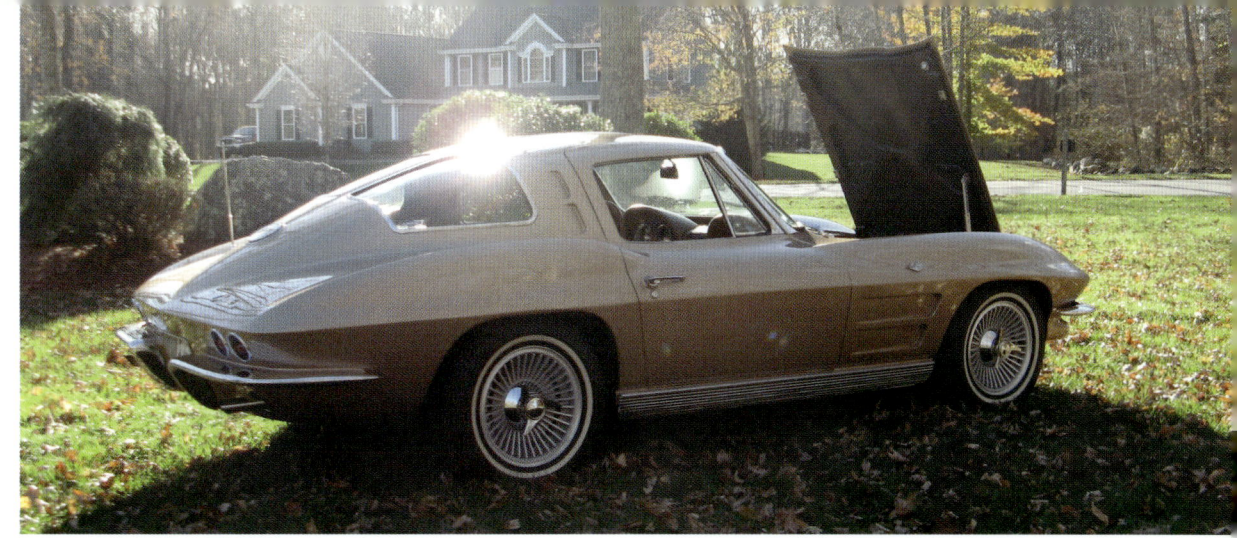

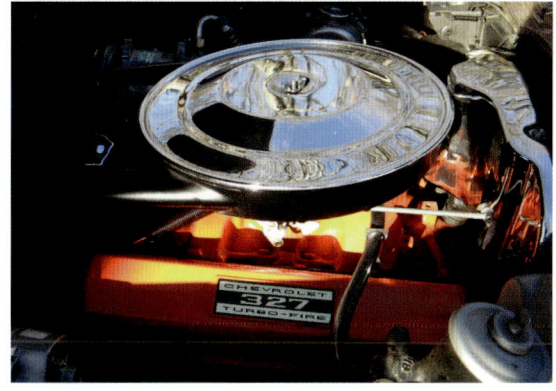

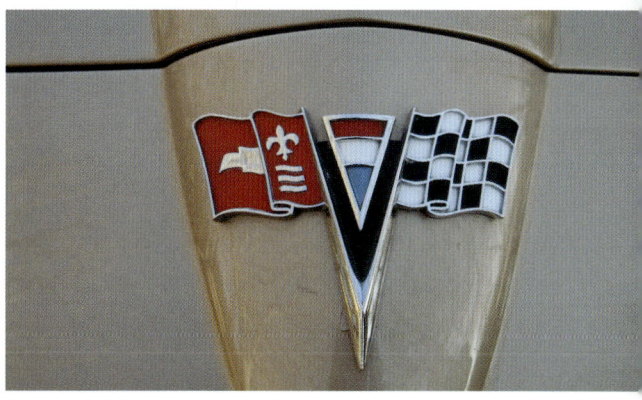

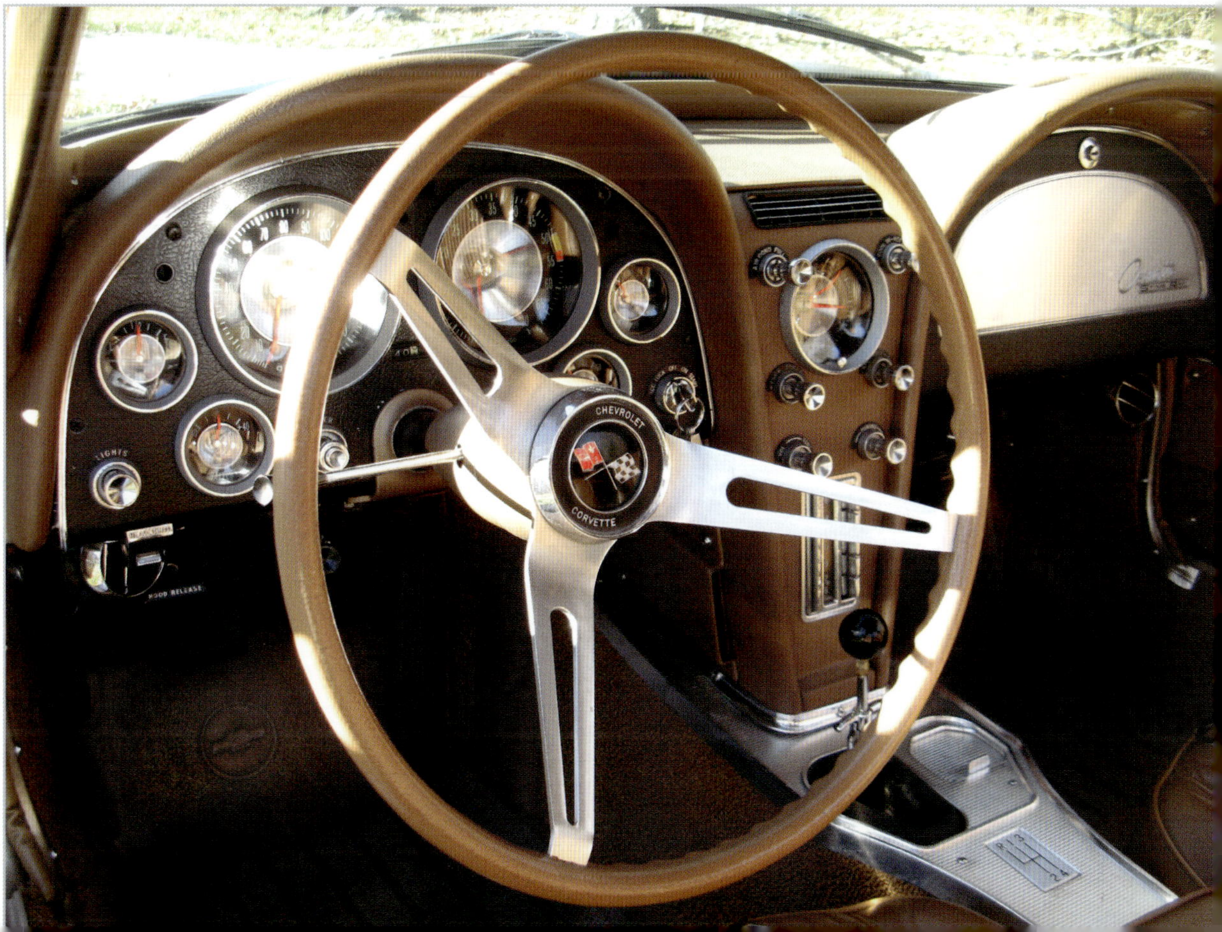
CHEVROLET
CORVETTE
LIGHTS
HOOD RELEASE

A Former Porsche Owner

2009, Red

Owner: Richard

Richard is a man of few words. However, when asked about this 2009 red Corvette, he answered, "After previously owning seven Porsches over the years, I finally bought the *true* American sports car!"

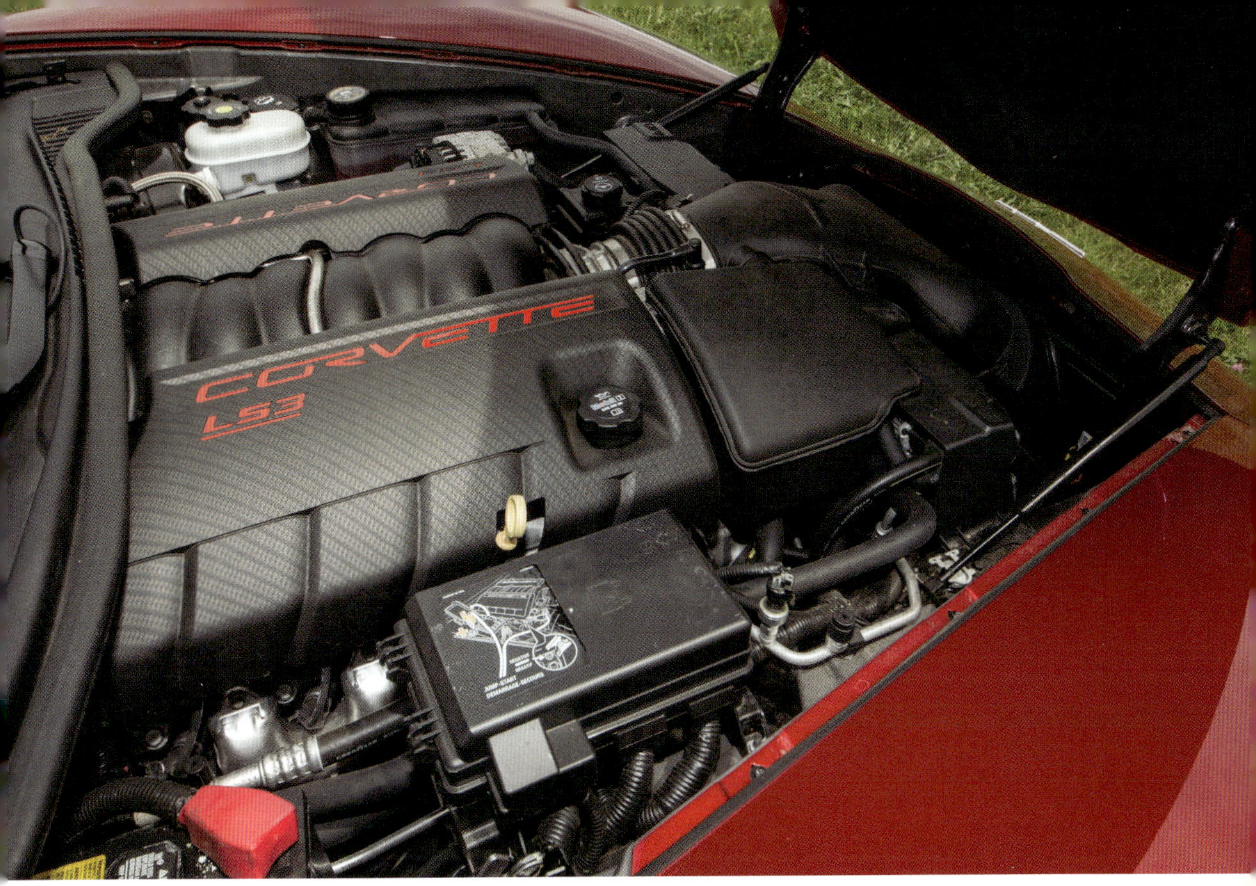
CORVETTE
LS3

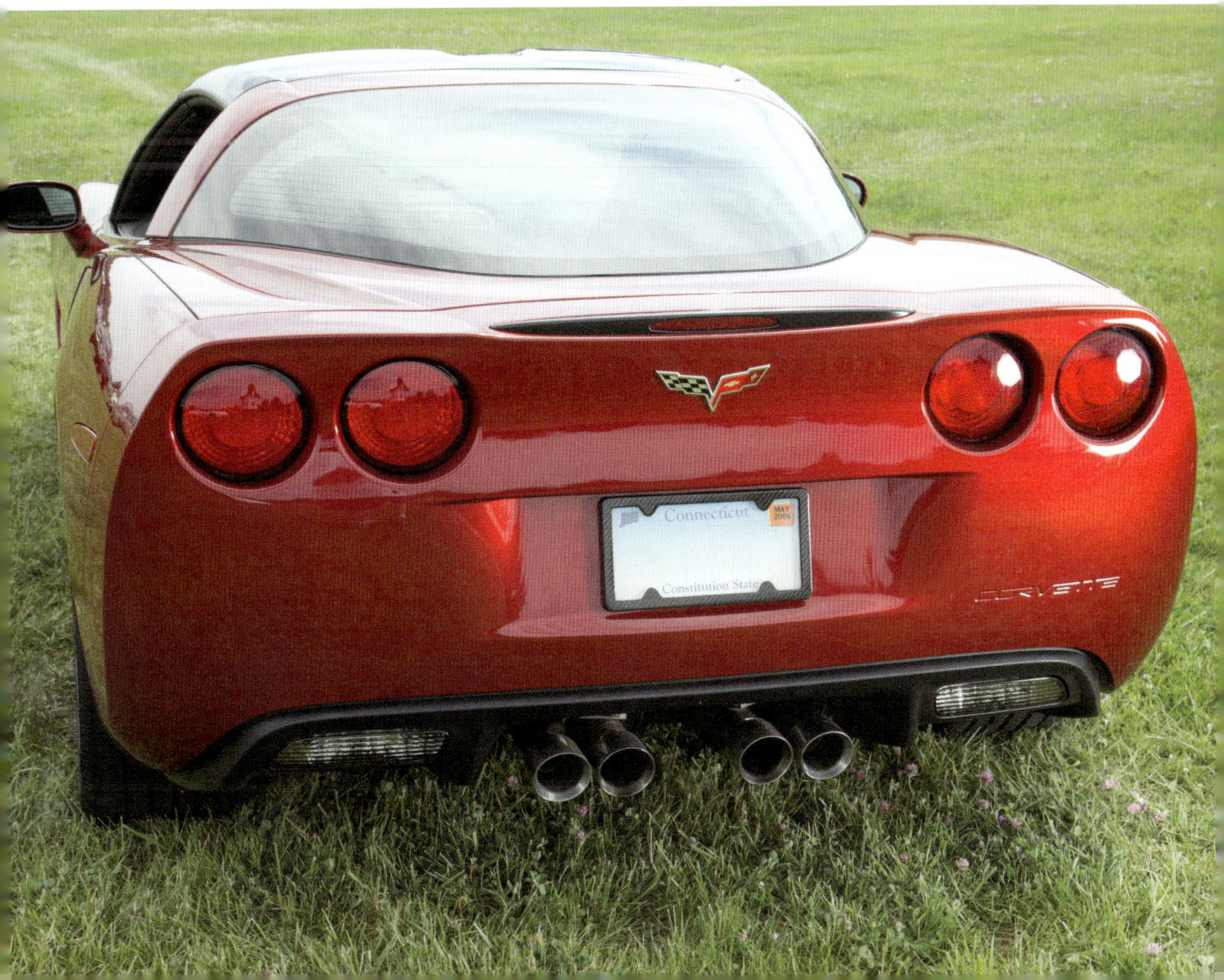
Connecticut
Constitution State
CORVETTE

A Dream Fulfilled

2010 Grand Sport, Torch Red
Owner: Bob

Since the early 1960s, Bob had wanted a red four-speed Corvette convertible.

In 2014 he spotted this 2010 Torch Red Corvette Grand Sport in a Chevy showroom. The four-speed was now a six-speed, but the rest was there.

"Beyond my original requirements, this car had big, wide tires, traction control, heads-up display, a big sound system and a *lot* of horsepower. I'm a Ford guy down deep, but I love this car!" Bob raves.

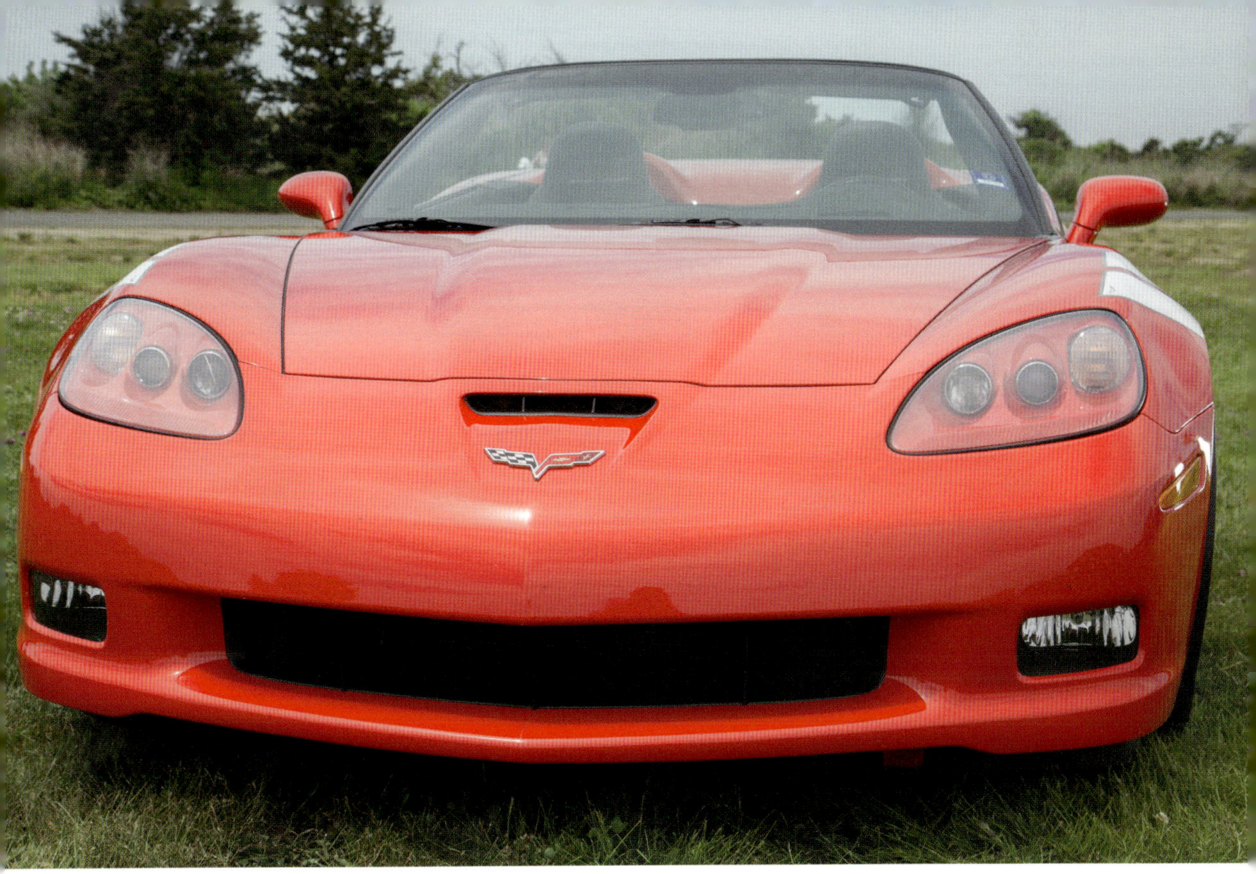

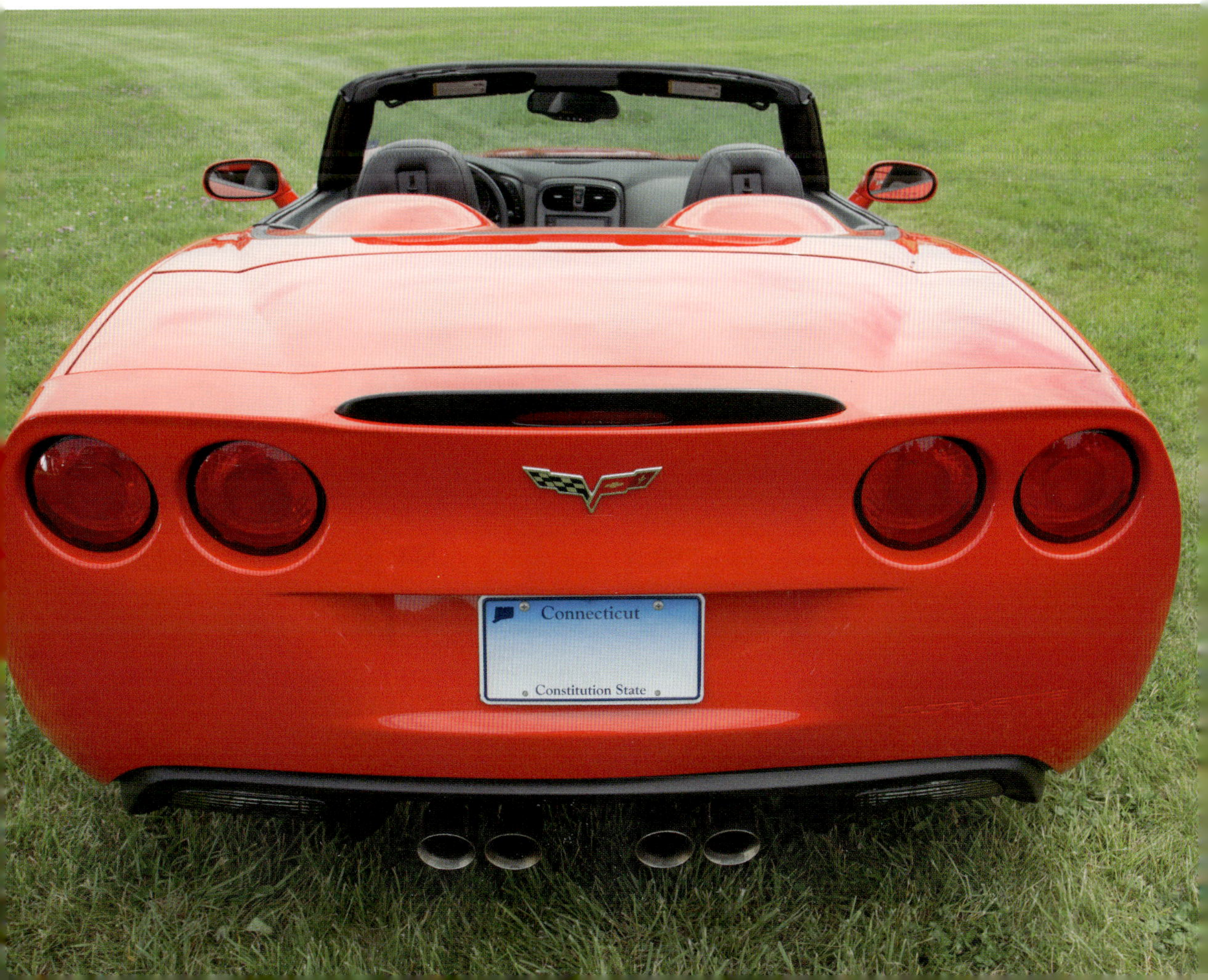
Connecticut
Constitution State

Quite a Learning Experience

1971 Convertible, Sunflower Yellow
Owner: Bob

When Bob purchased this 1971 Sunflower Yellow convertible in 2000, it was in pretty rough shape. He initially thought it only needed a quick paint job, but he was mistaken; it turned out to be the perfect example of project creep. "One thing led to another and there is not a thing that I haven't touched in some way," he says.

The car was restored and modified to be driven—and all of the work was done by Bob himself, including body work, paint, and interior. "Having never done this before made it quite a learning experience. Some of the modifications I made to make it a better driver were adding a GM ZZ4 crate engine, a five-speed transmission, a Borgeson power steering system, and a touring suspension system," he explains.

Bob has driven his rejuvenated Corvette throughout New England and as far south as Maryland—and he plans to continue driving it for many more years.

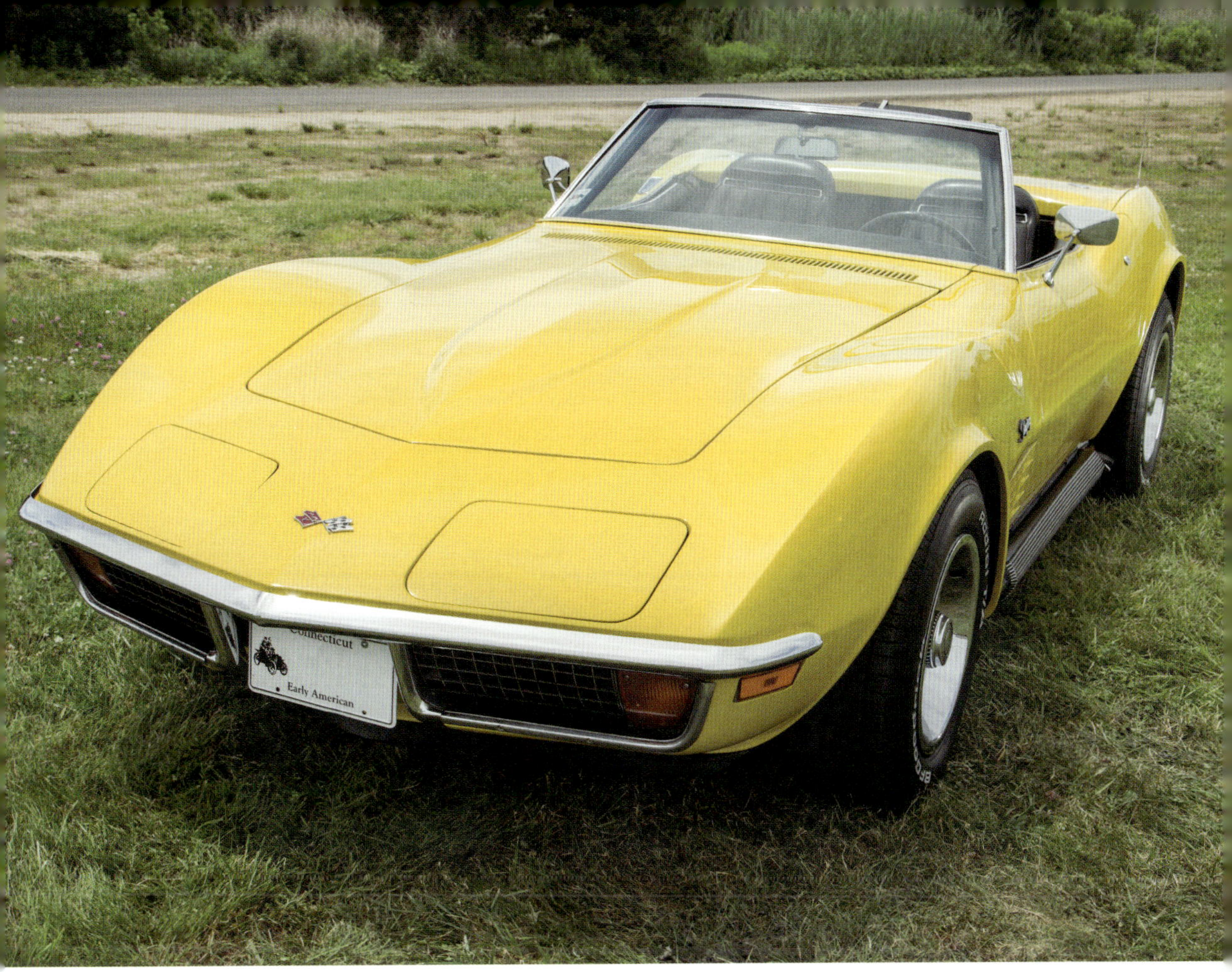
Connecticut
Early American

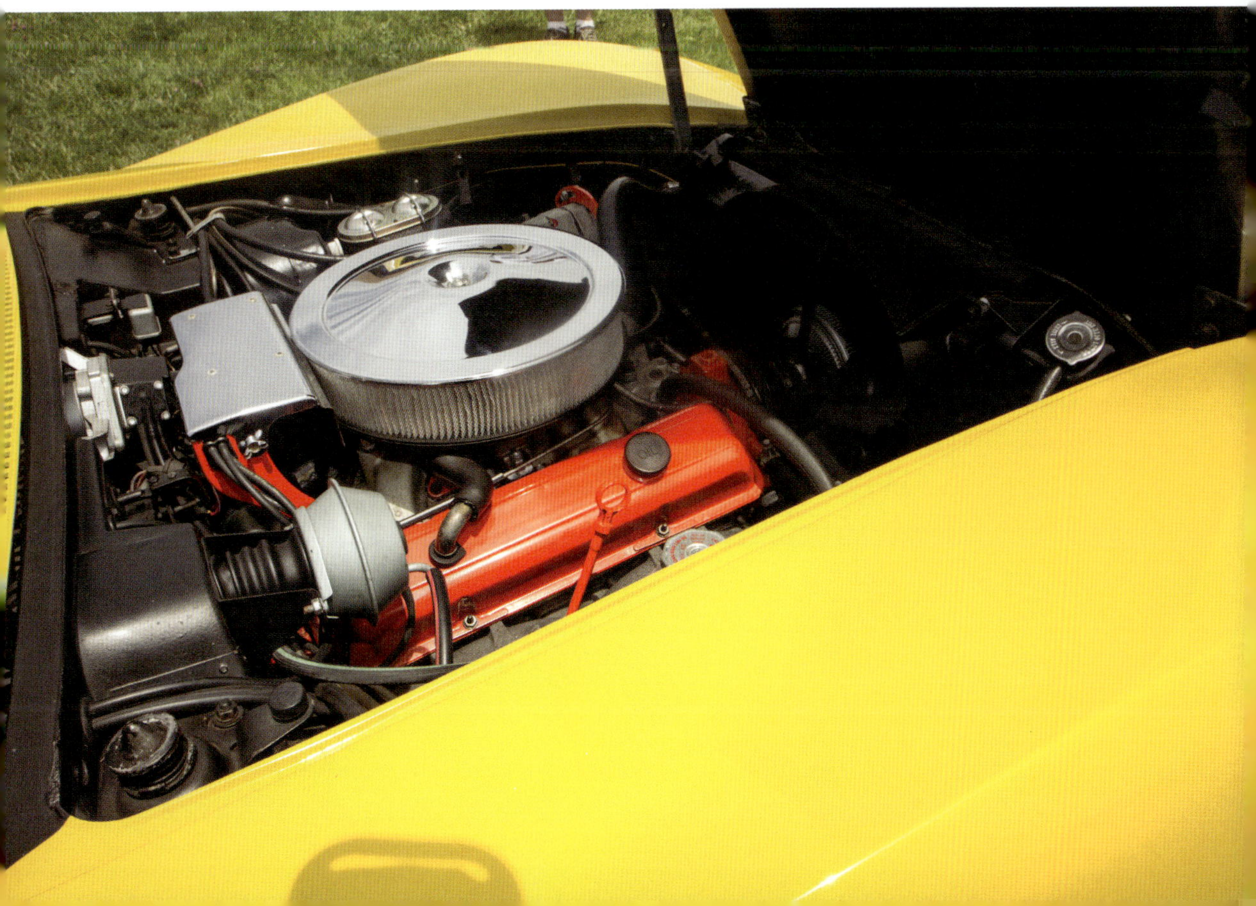

A Brother's Love

1996 Grand Sport, Admiral Blue
Owner: Michael

Michael owns a 1996 Corvette Grand Sport that is Admiral Blue with an Artic White stripe. His love for Corvettes began in the summer of 1957 when his cousin's boyfriend had an intimidating black 1957 "fuelie." Every time he went through the gears, Michael was listening—and it was all he thought about.

"A few years later I bought a '55 Chevy Bel Air," he says. "I drove that beauty for two years, never forgetting that '57 Corvette. I came across a 1957 Corvette with a 270 motor, but I did not have enough money to buy it. My brother, who was sixteen years older and raised me, was also a car lover and unbeknownst to me, cut a deal with the owner."

"I came home one night after work, opened the door to the garage, and there was

my dream car!" Michael continues. "I must have sat in that car for two hours! The next morning my brother woke me, saying that his car would not start and he needed my help. I had to perform the ultimate act of surprise. When we opened the garage door, there was the black beauty! My brother and I hugged and cried with joy."

"I never told my brother that I had known about the car, because it made him so happy to have surprised me," Michael says. "When I was at my brother's bedside on his ninetieth birthday, he mentioned how excited I had been when I saw that 1957 Corvette. He passed away one week later, never knowing that I knew what he had done."

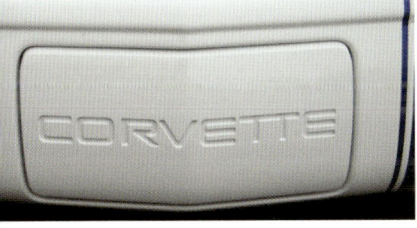

Their Fourth Corvette

2011 Convertible, Cyber Gray
Owners: Madeline and Anthony

Many Corvette owners have purchased more than one over the years; Madeline and Anthony have owned four.

Their first Corvette was a 1962 beige convertible; this was followed by a 1968 blue Corvette. The years passed, and with a growing family, they did not get another Corvette until 2013 when they purchased their son-in-law's beautiful 1990 Polo Green convertible.

Their son-in-law had not used the car for years, so after Anthony bought it, he put over $5,000 into the car to get it back in condition. He kept it for a few years, but he was not fond of the C4 style because it was difficult to get in and out of the car.

In 2011, they purchased this new Cyber Gray Corvette and they love it!

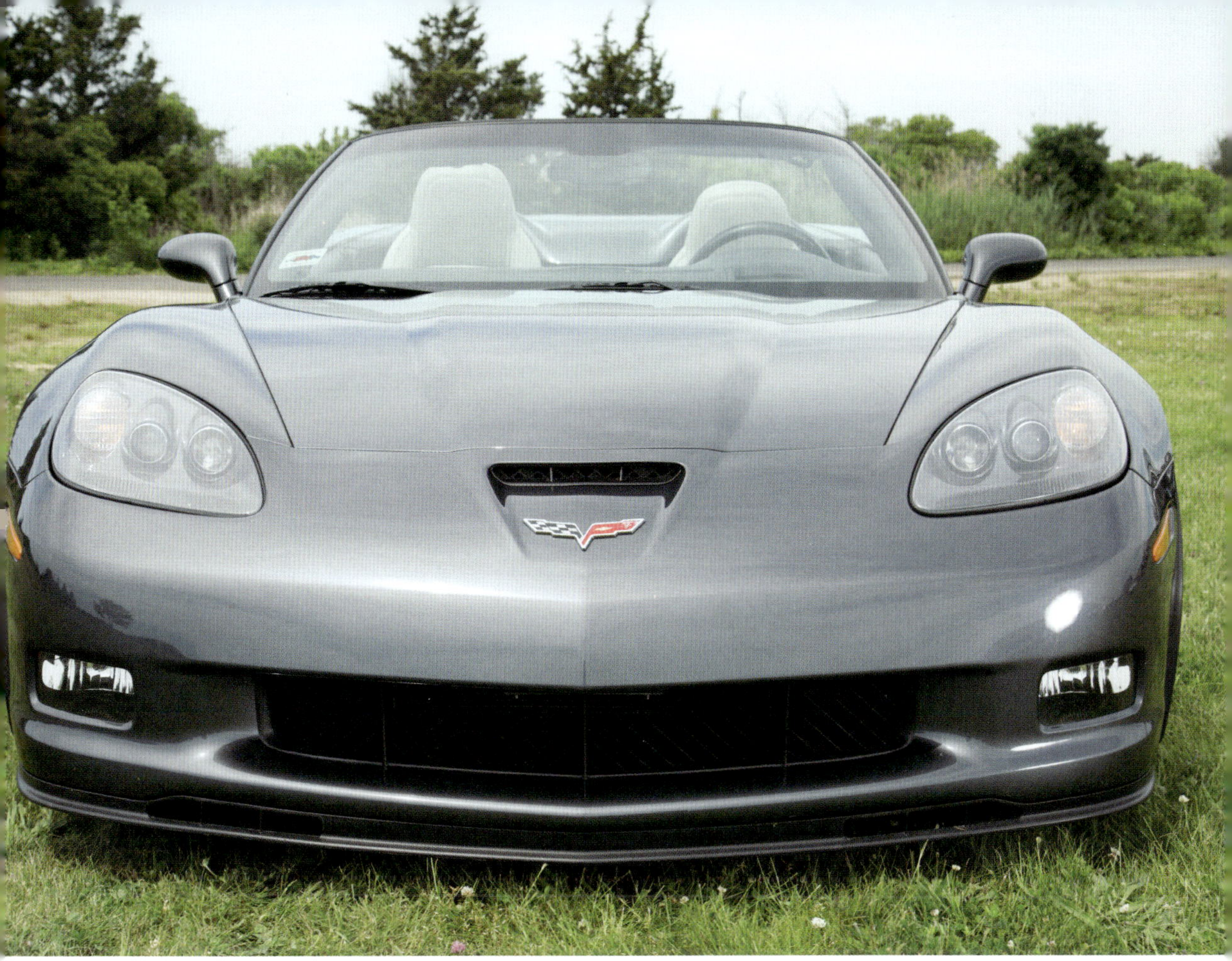

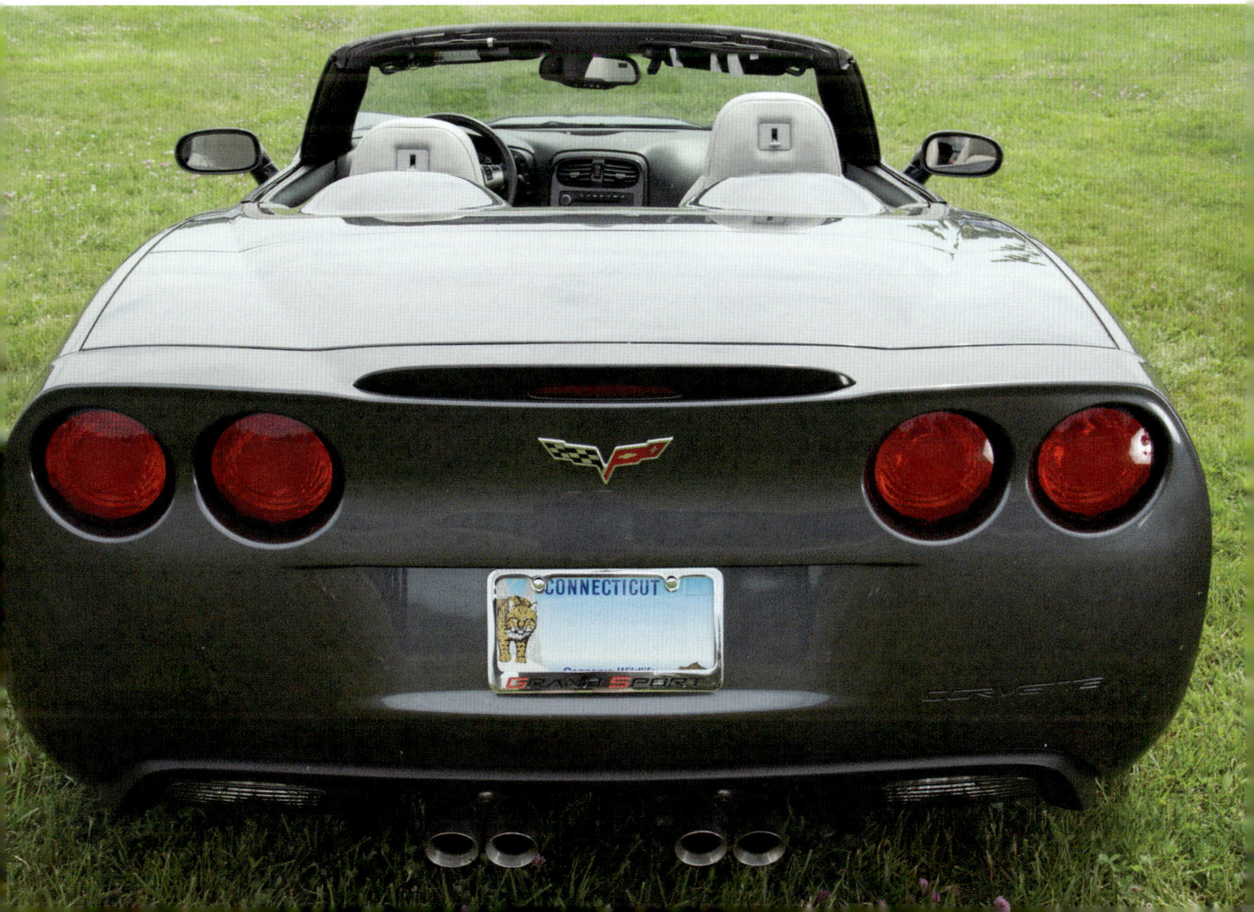
CONNECTICUT

A Second Childhood

2000 Convertible, Magnetic Red

Owner: Ken

Ken's love for Corvettes began in the summer of 1966 when he worked at the neighborhood grocery store. Every Saturday, around closing time, the manager would throw Ken the keys to his 1953 Corvette so he could bring it around to the front of the store.

"That was it, I was hooked—and just like every young man in the United States, I dreamed of owning a Corvette of my own.

Unfortunately, life got in my way. College, marriage, and a family occupied all my time," says Ken. "It wasn't until I entered my second childhood that I found this 2000 Magnetic Red Corvette. I get the same feeling driving it today as I did with that '53 all those years ago. This is my first Corvette, but she won't be my last."

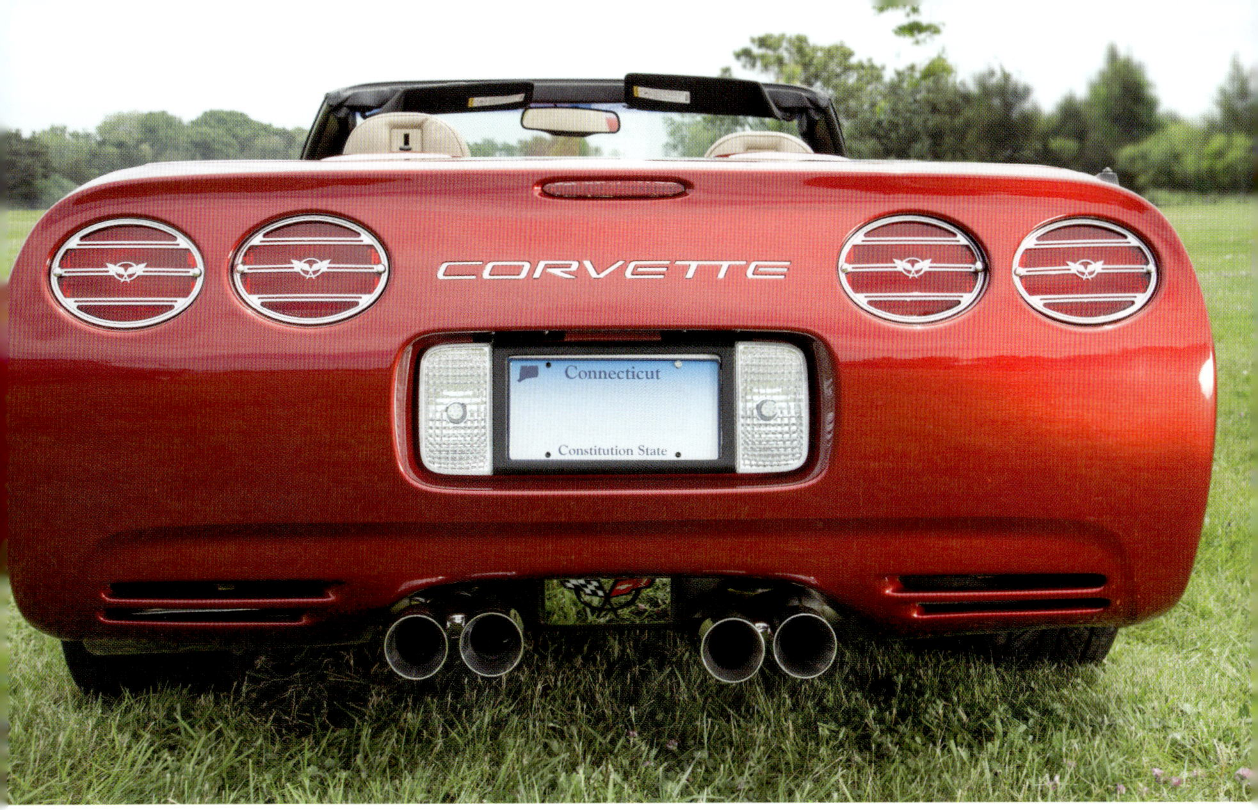
CORVETTE
Connecticut
Constitution State

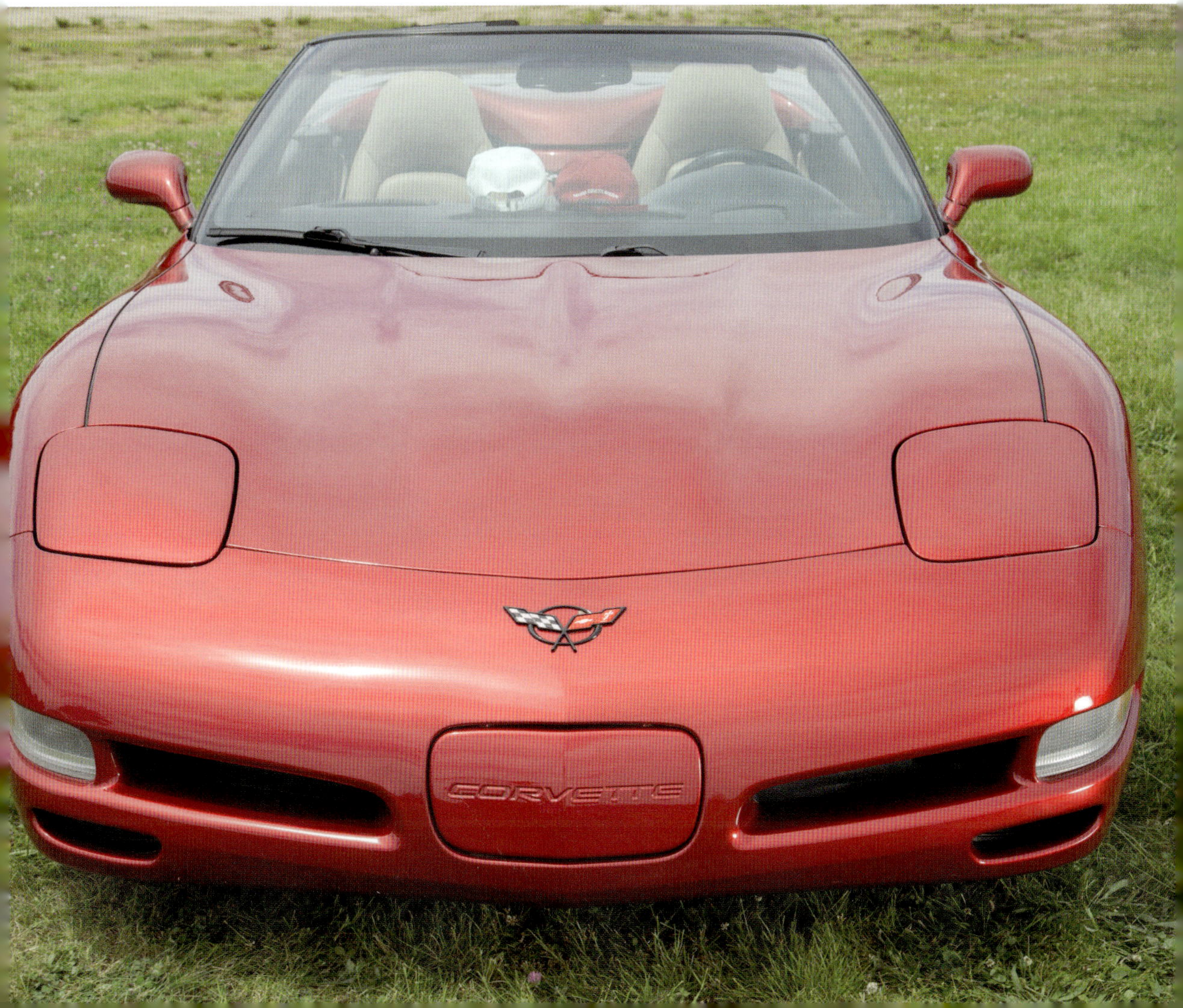
CORVETTE

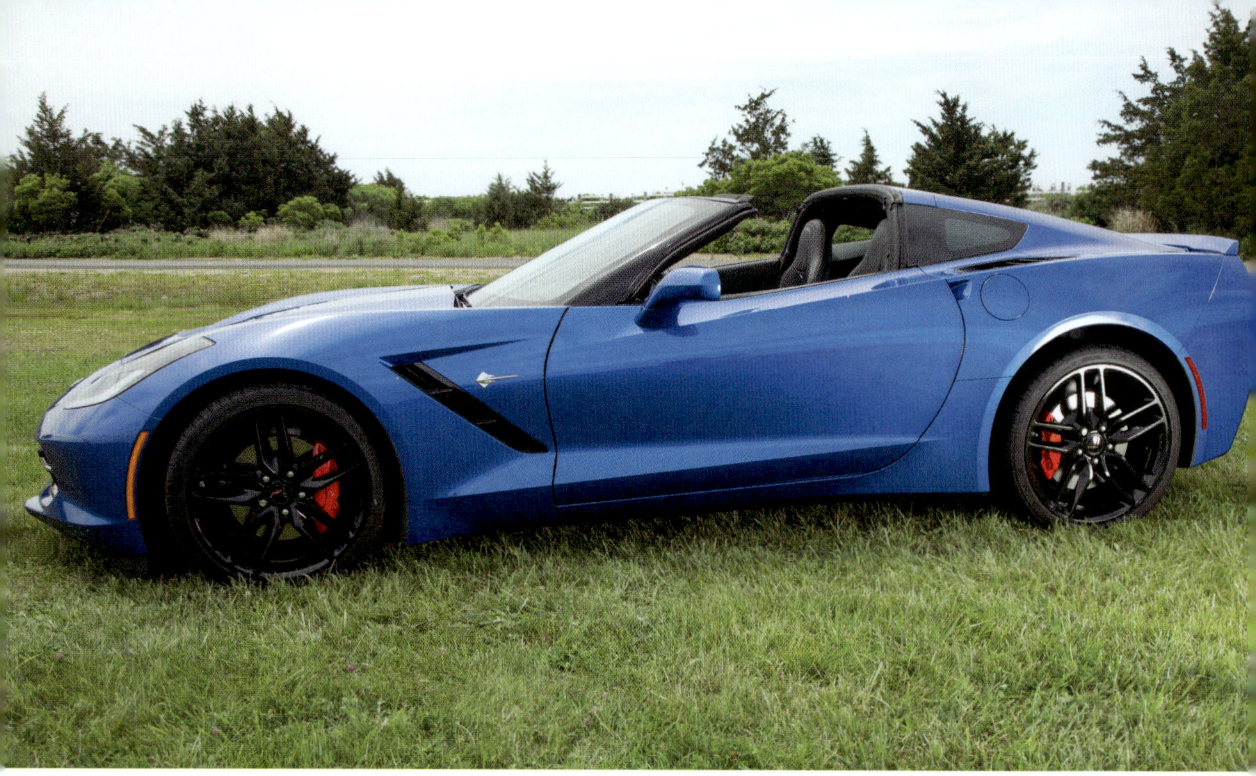

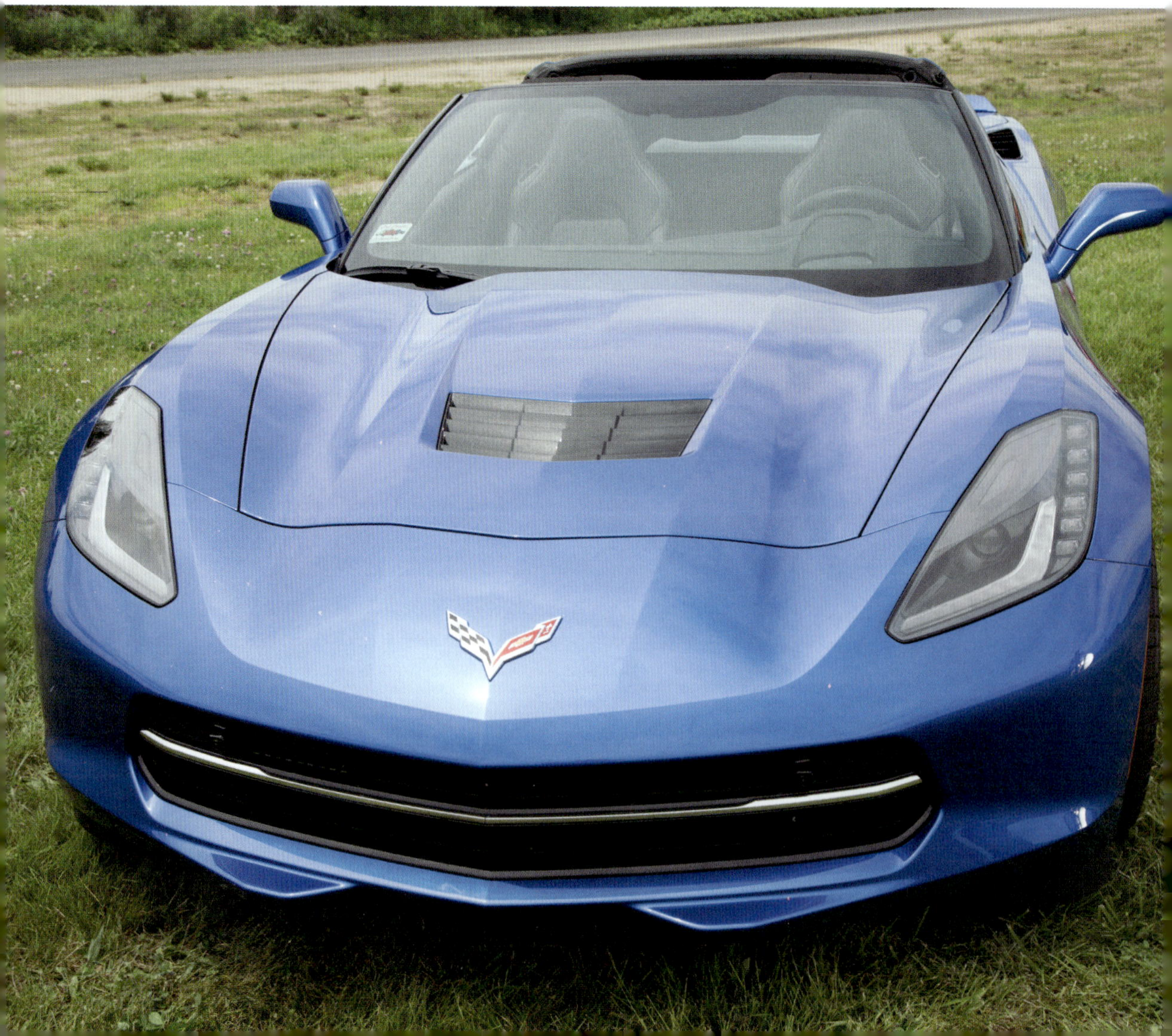

The Best of Everything

2016 Stingray LTII, Blue
Owner: Tom

Tom is not new to Corvettes, but this 2016 blue Corvette Stingray LTII offers him the best of everything. "This is by far the most technologically equipped, best riding, most comfortable Corvette I have ever owned," he sayd. "Like any Corvette, getting in and out for someone like me that has had major back surgery is always a challenge. However, once you are in that driver's or passenger seat, it feels like a glove is wrapped around you."

"The car gets fantastic gas mileage for this type of vehicle. On the highway in ECO mode, I have gotten as much as 28 to 29 miles to the gallon," he adds. "While I have been known to change cars frequently, this Corvette will be with me for some time to come!"

Still Cruising

1966, Mosport Green
Owner: Forrest

Forrest's first Corvette was a 1965 Glen Green convertible with Saddle interior, purchased in 1970. It was fun for rallying, autocrossing and general weekend trips—but when his second child arrived, the convertible had to go.

His second Corvette was a 1967 Sebring Silver coupe with black interior. He and his wife still enjoyed rallying and weekend trips—now with the kids rolling around in the back. They added a fresh coat of paint and 1967 bolt-on wheels with red line tires and side pipes. By 1978, however, the new Corvettes had cast a spell on him.

"A 1978 beige Corvette LT-1 with deep brown interior and T-tops then took the space in our garage. Again, we rallied and auto-crossed, but something was lacking. It didn't sound like a Corvette and it drove like a Buick with power this and power that—and, of all things, air conditioning. It wasn't our idea of a Corvette."

"With our kids starting families of their own, we began looking for a sports car. We decided to go with something we knew—something fun, something to turn back the clock. We began looking for a Corvette from 1965 to 1967. While I preferred a '65 or '67, my wife loved the 1966 Mosport Green. We found a rust-free, two-top convertible for sale in Arizona. It had originally been purchased in Texas and spent all its time in dry climates."

They no longer autocross or drag race, but the couple still delights in cruises, small shows, and weekend drives in their 1966 Corvette!

A Chance to Live His Dream

2001 C5, Pewter
Owner: Linda

Linda's husband died at age 58 of Lou Gehrig's Disease. Besides his family and friends, the two things that he loved most were the New York Yankees and Corvettes.

He wanted a pewter C5 Generation Magnet and always said, "Someday, when we retire . . ." Unfortunately, he never made it to that day. But approximately six months after he passed away, a 2001 C5 Pewter Corvette popped up on eBay in Jamaica, Queens—where he was born—and Linda bought it!

"It was as though it spiritually popped up on eBay for me to see and live his dream," she says.

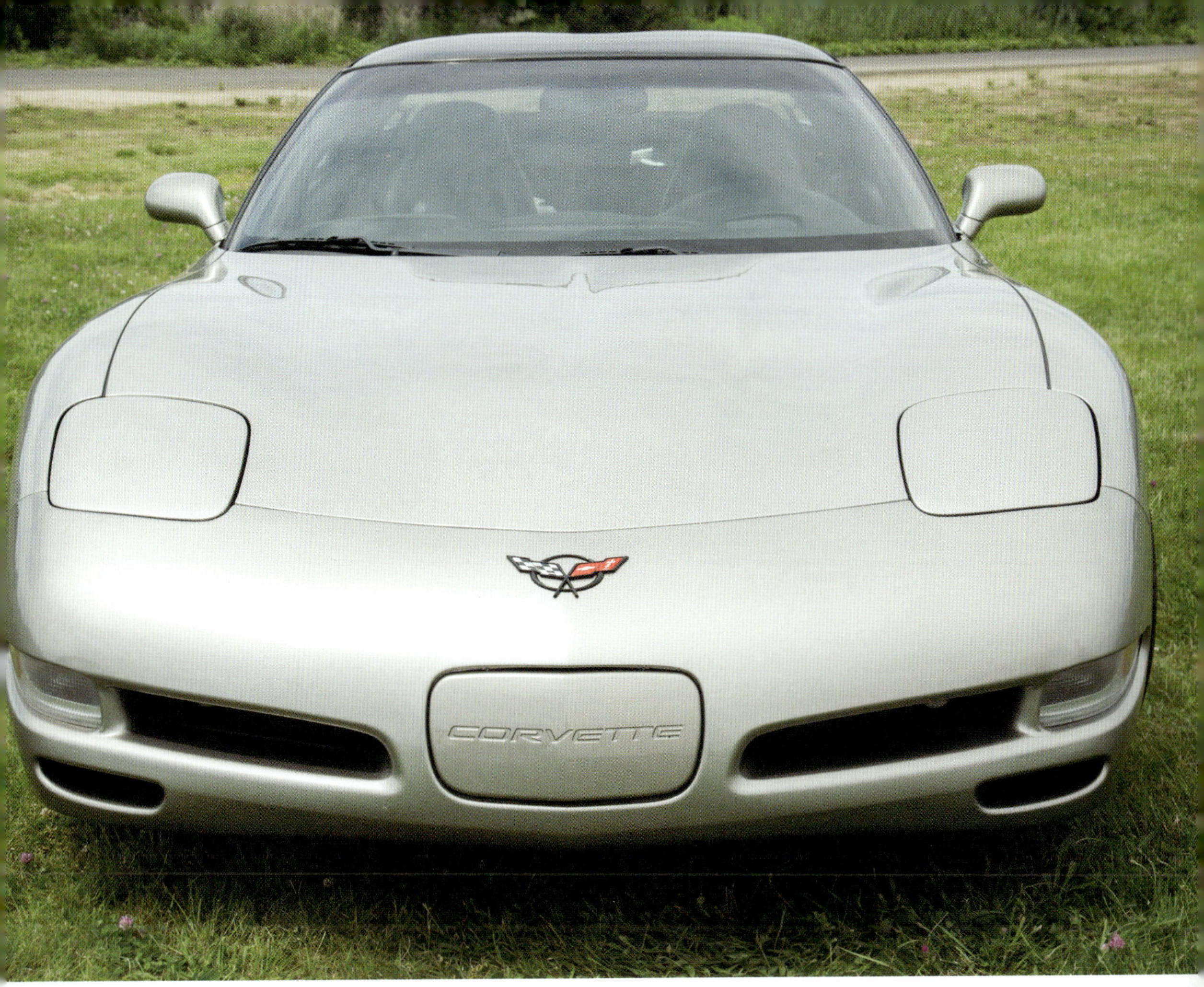
CORVETTE

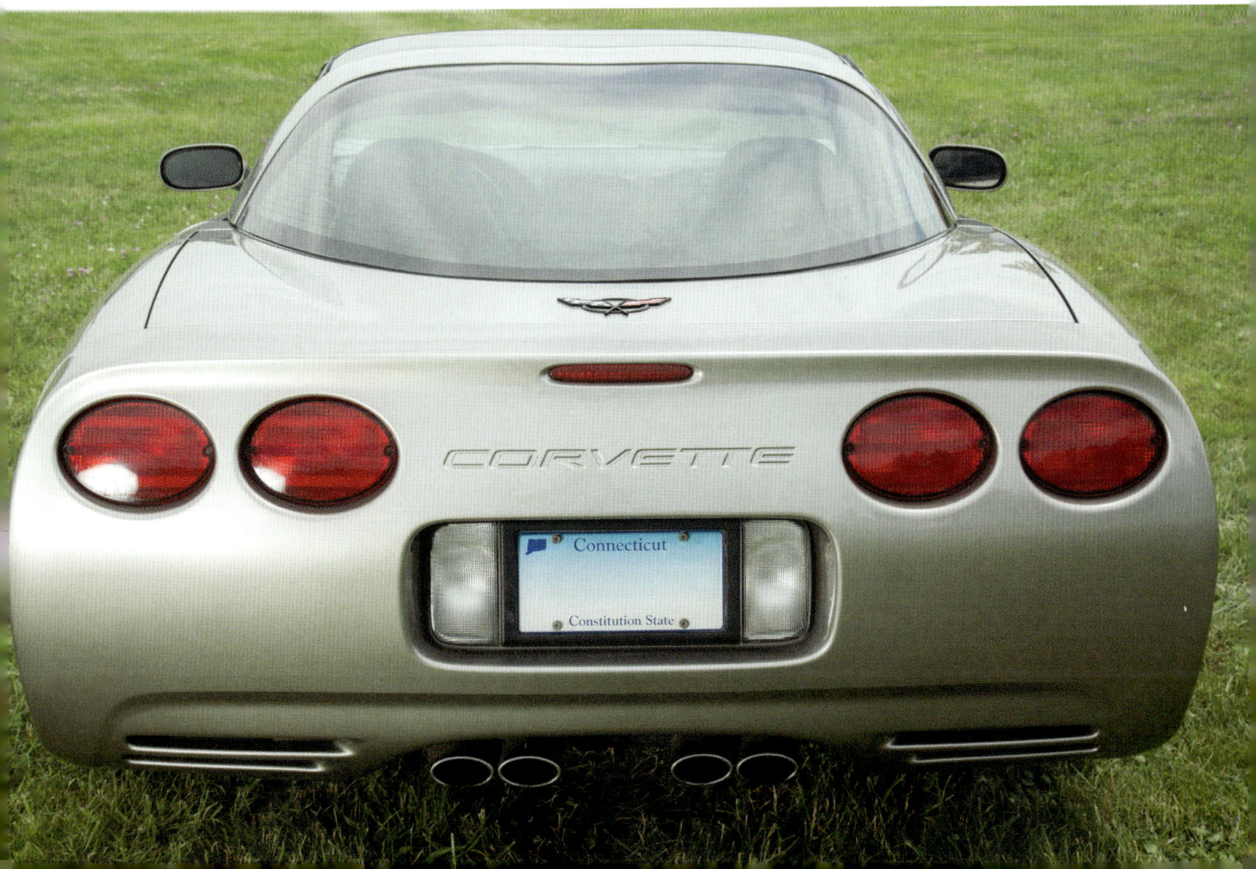
CORVETTE
Connecticut
Constitution State

A Good Fit

2015 Stingray C7 Convertible, Torch Red
Owner: Jack

The 2015 Stingray C7 convertible was the first Corvette that Jack was able to fit in comfortably. He is over six feet tall, so the previous models never worked.

"Online, I just fell in love with the convertible with chrome wheels, Torch Red exterior, and black leather interior," he says. "I ordered customized accessories in July and by August 20th the car left the production line. I received delivery on September 4, 2014. It was the first 2015 C7 that was delivered to Connecticut."

In June of 2016, Jack drove to Indianapolis to attend the Bloomington Gold, where thousands of Corvette owners from all over the country gathered at the world famous Indianapolis Motor Speedway to participate in the three-day event that included Gold Certification, Autocross, Chevy Ride & Drive, laps around the IMS oval, Gold Tour, and more. The highlight for him was driving laps on the Indianapolis Motor Speedway.

A Father-and-Sons Project

1966 Big Block, Nassau Blue
Owner: Steve

Steve's passion for automobiles began as a teenager when he started helping his dad work on cars in their garage. Steve got his driver's license in 1966 and spent most of his weekends at the local McDonald's, watching other guys drive through in their Corvettes. In 2002, he had the opportunity to buy a numbers-matching 1966 big block Nassau Blue Corvette—just like the ones he had admired as a teenager. Between 2004 and 2008, Steve (along with his dad and his brother Mike) worked with a local restoration shop and completed a full restoration on the car. This was the brothers' last father-and-sons project before their dad passed away. The Corvette holds special memories for Steve and his brother, who enjoy showing the car at cruise nights and car shows.

Off to See America

2017 Grand Sport, Long Beach Red
Owner: Rod and Linda

Rod and Linda love to travel—and what better way than in a 2017 Long Beach Red Corvette Grand Sport? They have climbed Pikes Peak, Mount Washington, and Cadillac Mountain—and adventured from Nova Scotia to Key West. All along the way, they did the Corvette Wave and heard, "Beautiful car!"

Rod says, "We have done laps on race tracks and actually drove 140mph across the country on Route 70. We did the Strip in Las Vegas, the Four Corners of America, and Route 66. Our next adventure will be circumnavigating the Great Lakes. There is no better way to see America."

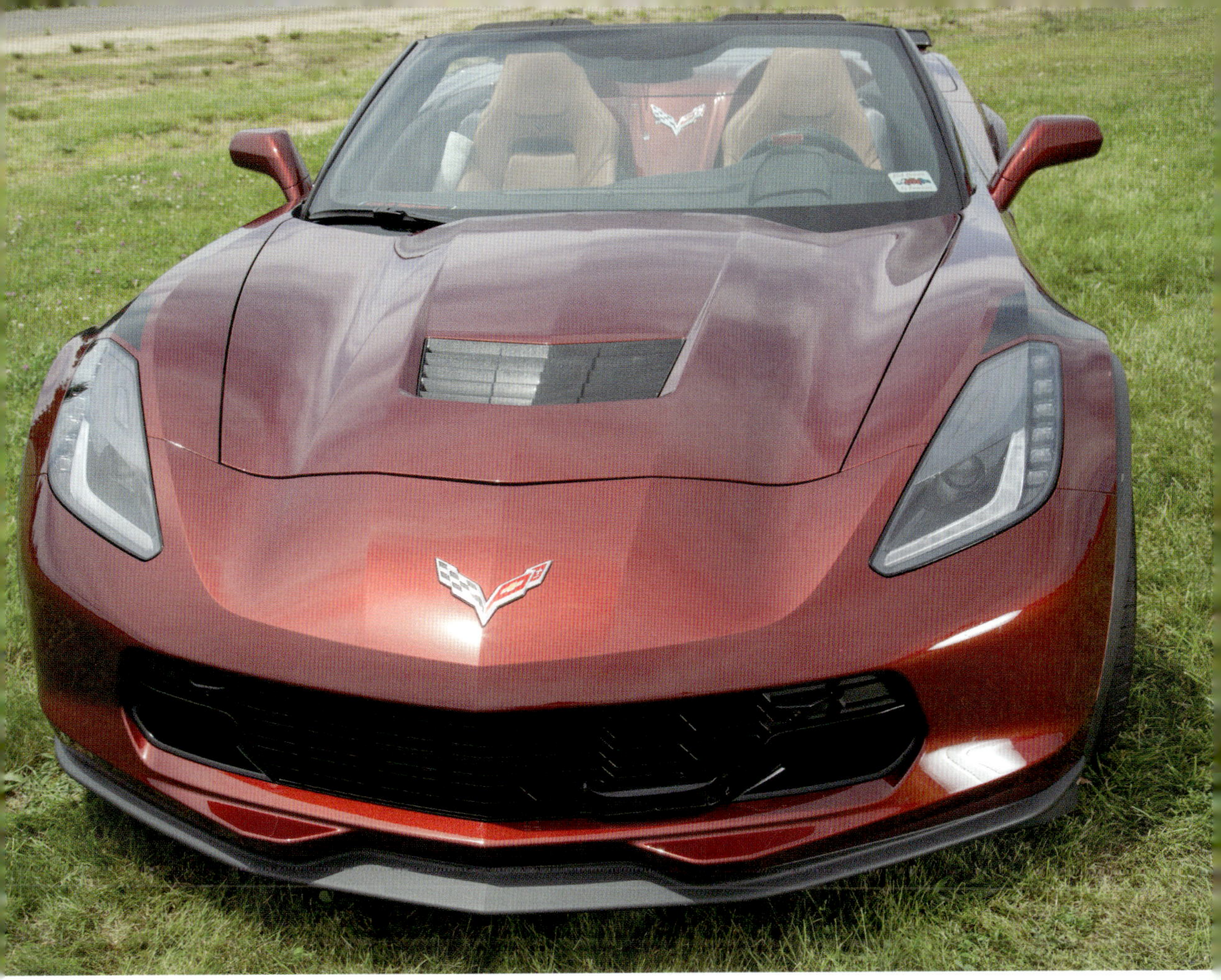

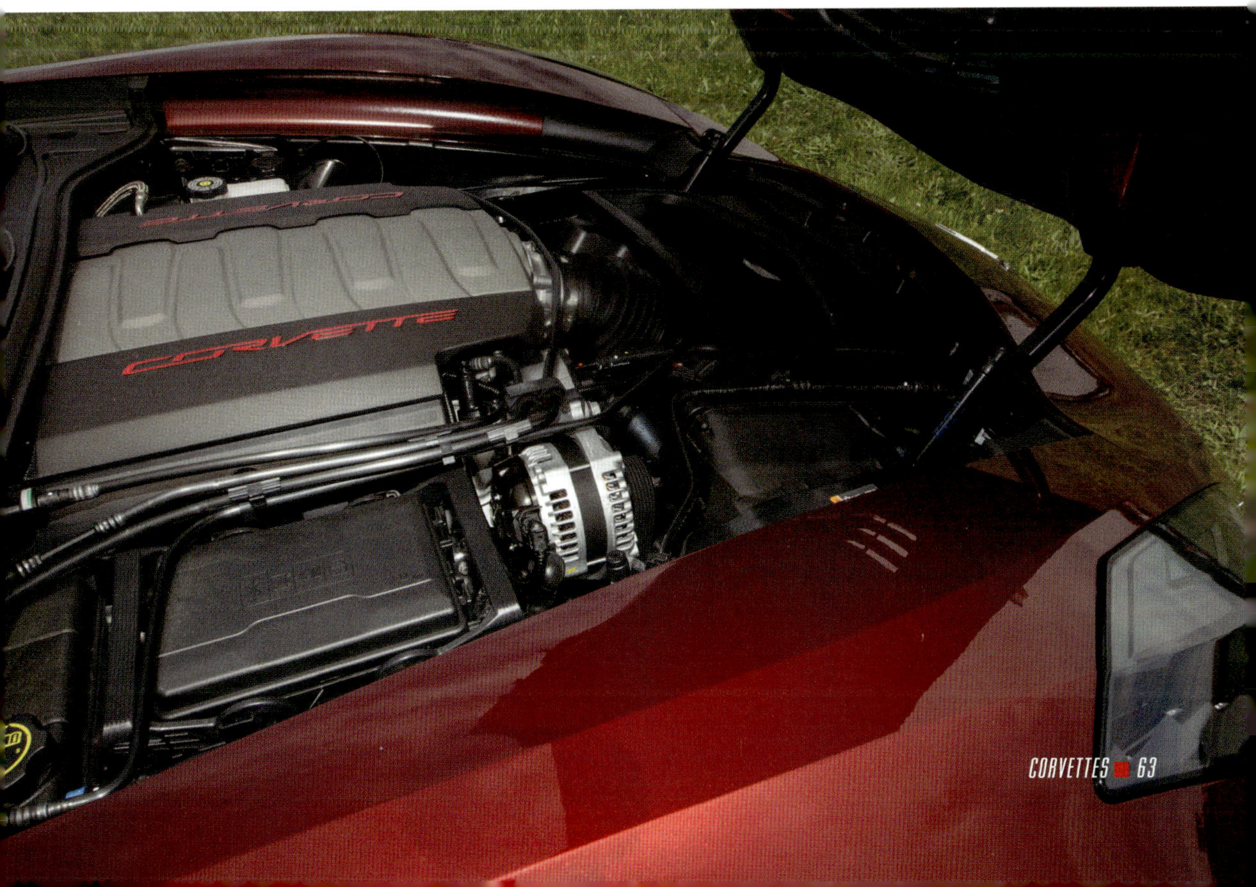
CORVETTE
CORVETTE

Miles and Smiles

1999 Convertible, Pewter

Owner: Richard

This is one of Richard's two Corvettes; he also owns a 1988 model. "I purchased the 1988 from a friend and the 1999 pewter Corvette from a dealership in Plymouth, Massachu- setts. Both Corvettes have given me many hours of enjoyment on the road and swapping stories with other Corvette owners," says Richard.

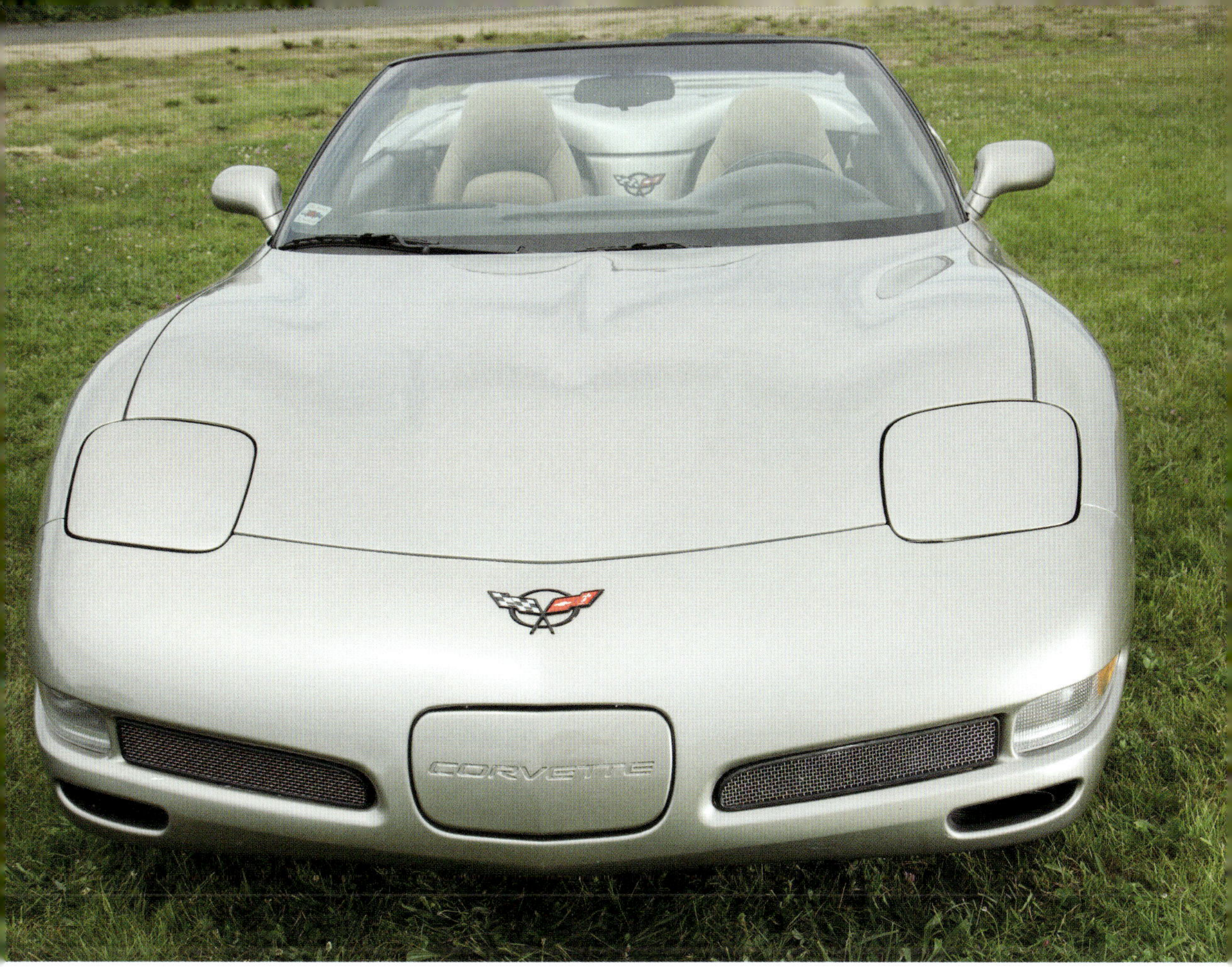
CORVETTE

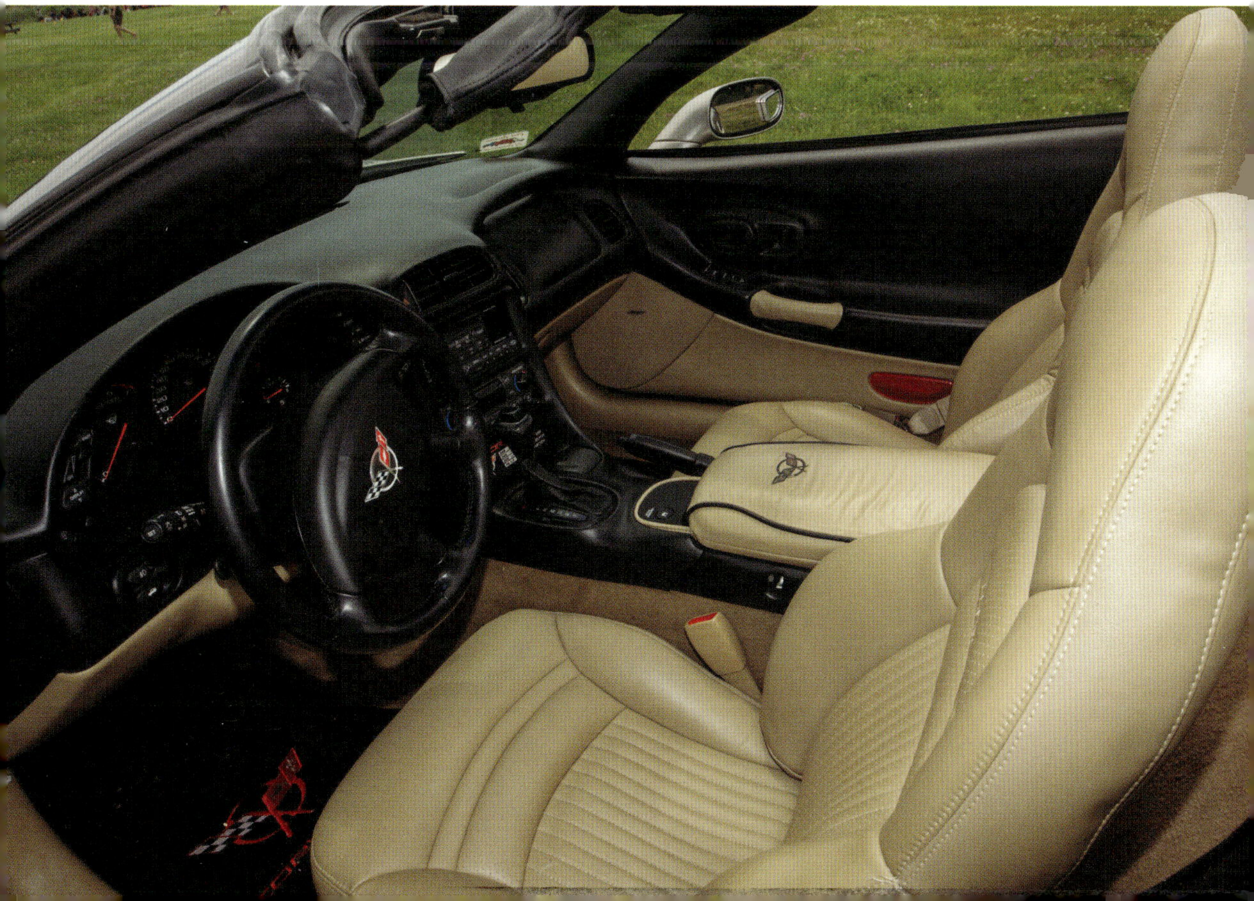

Rebuilt and Restored

1958 Convertible, Red

Owner: Don

In 1976, Don flew from New England to Akron, Ohio, to purchase this red 1958 Corvette and drove it home. "The car was pretty ratty except for the paint job," he says. "I did a body-on restoration in 1980 and rebuilt the motor, refurbished the interior, and replaced the chrome."

"I raced the car at Lime Rock in the late 1980s in vintage sports car events. I also took it to the Englishtown, New Jersey, dragstrip a few times—and to a lot of rallies. This Corvette is a 'driver;' I even take it out in the winter with the hardtop on, unless there is snow or salt on the road."

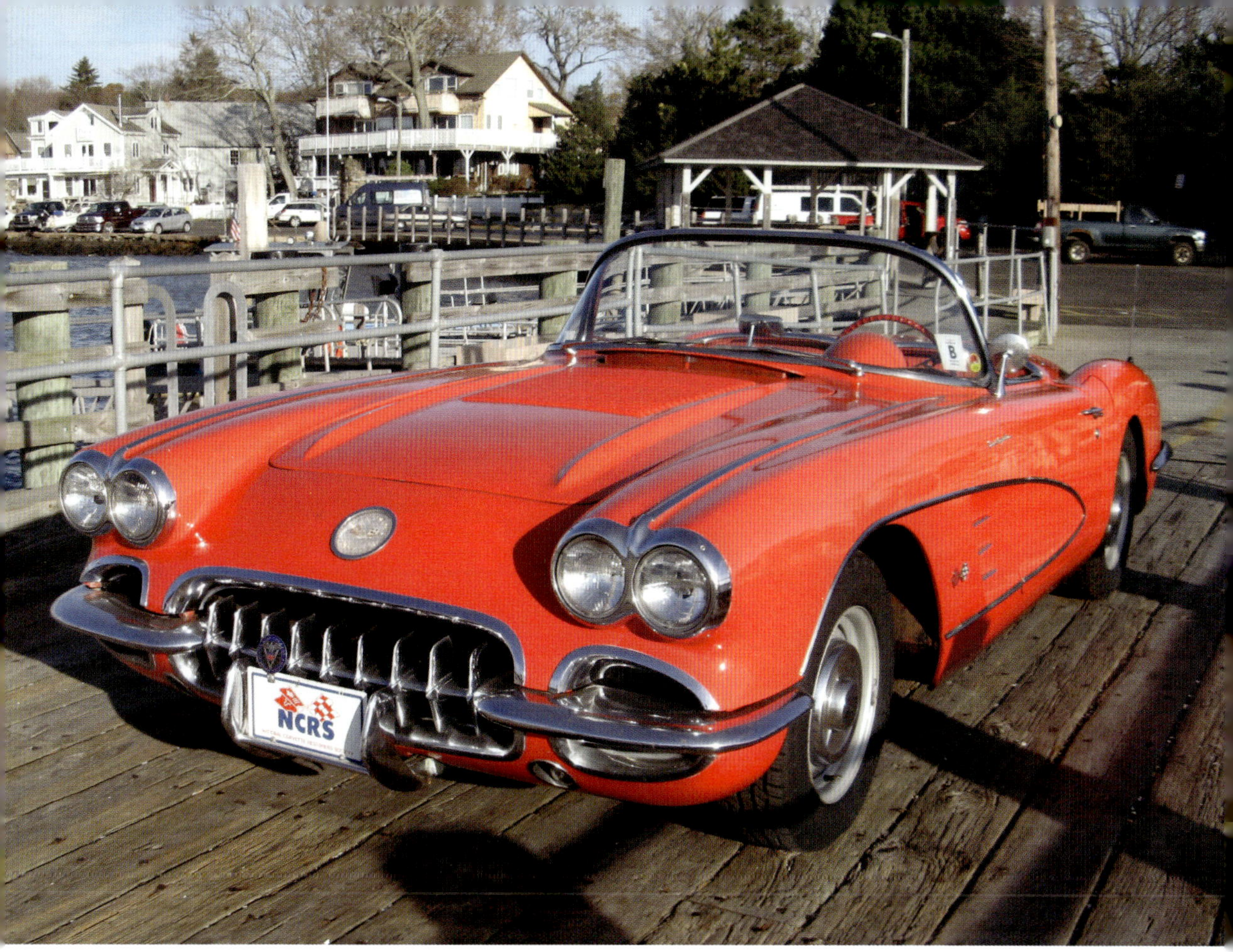
NCRS

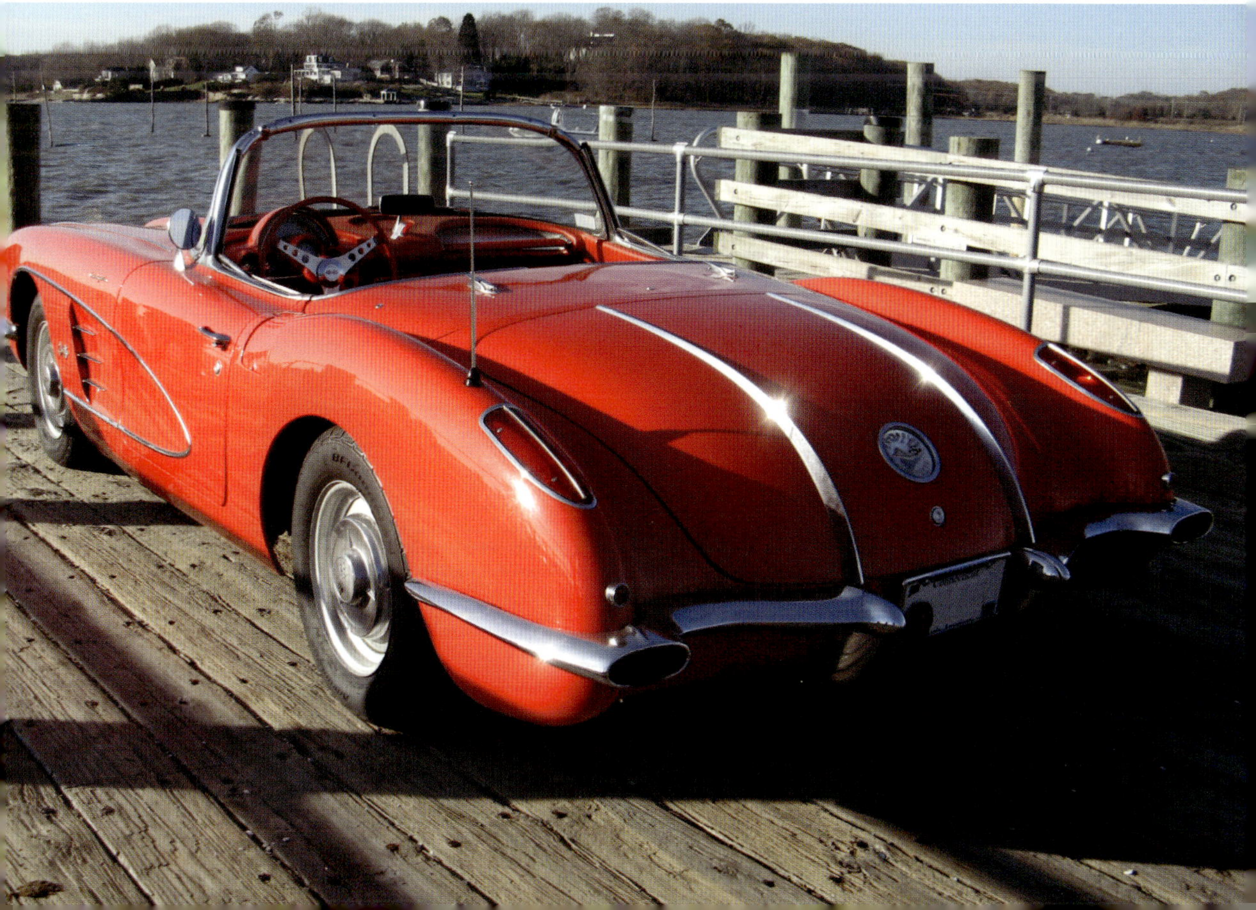

Life Is Short–Buy the Car

2016 Stringray Z51, White
Owner: Walter

Walter waited until 2011 to buy his first Corvette. "While I always liked Corvettes, the Mustang was my favorite car when I was young," he says. "It was not until my mid-life crisis that my appreciation for the 'American Sports Car' surpassed the Mustang. My first Corvette was a fully loaded 2007 Victory Red convertible with 14,000 miles."

However, when the C7 came out in 2014, Walter says it was love at first sight. "After a couple of years spent agonizing over whether or not to spend the money to upgrade. After countless nights on the computer visiting Chevrolet.com and the local dealers' inventory, the answer became clear: life is too short. On July 1, 2016, I became the proud owner of a white 2016 Corvette Stingray Z51 3LT."

One year and 2,300 miles later, Walter says he has absolutely no regrets—and adds, "I also don't miss hearing my wife scream about what the convertible did to her hair!"

Feed the Need for Speed

2011 Callaway Grand Sport, Metallic Orange

Owner: Richard

Richard loves his 2011 Callaway Corvette Grand Sport in Metallic Orange for its pickup and speed. "My 2011 Callaway Corvette Grand Sport has 600+ horsepower," says Richard, "and with its chrome supercharger, it can go as fast as 200 miles per hour—and from zero to 60 in 3.4 seconds."

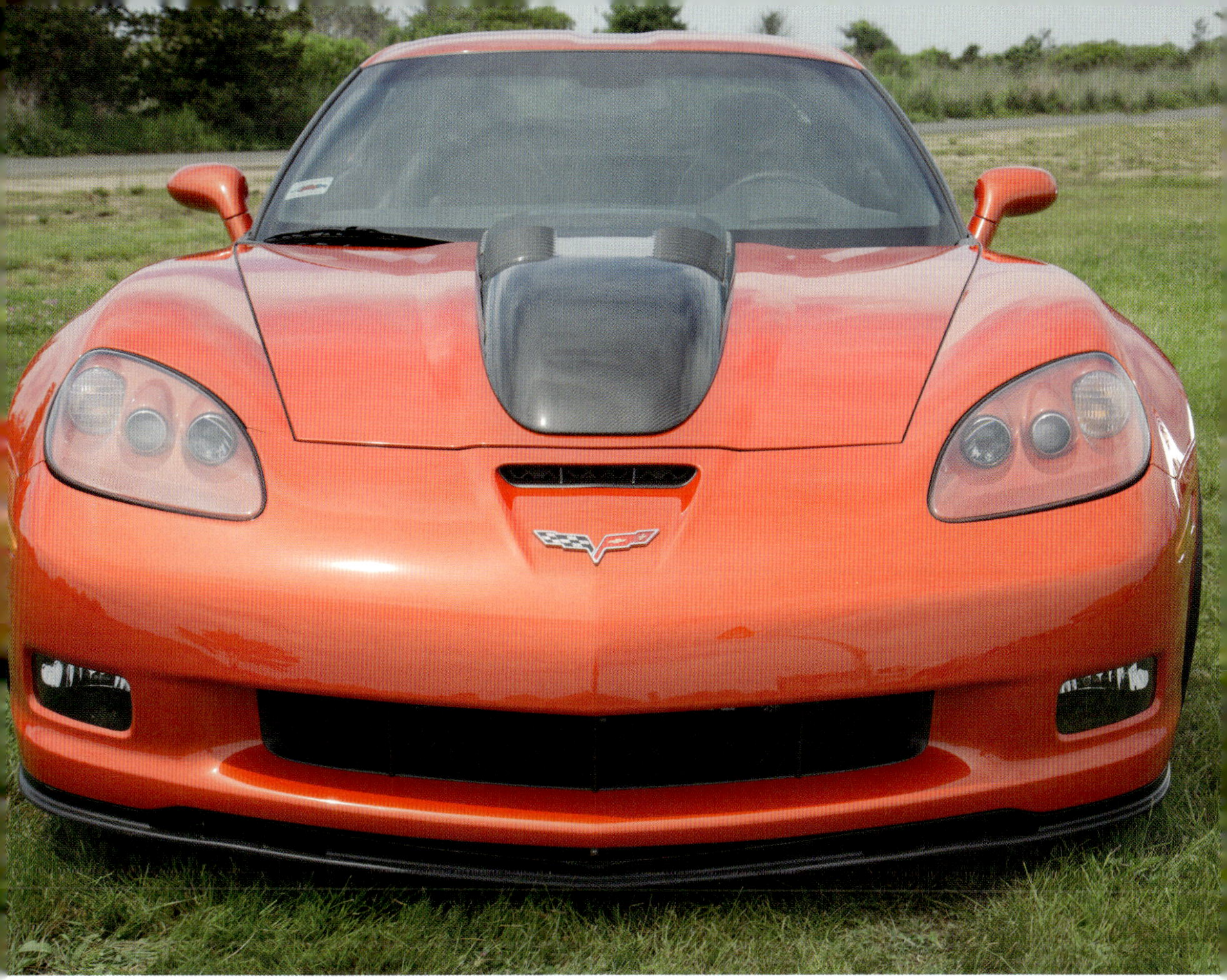

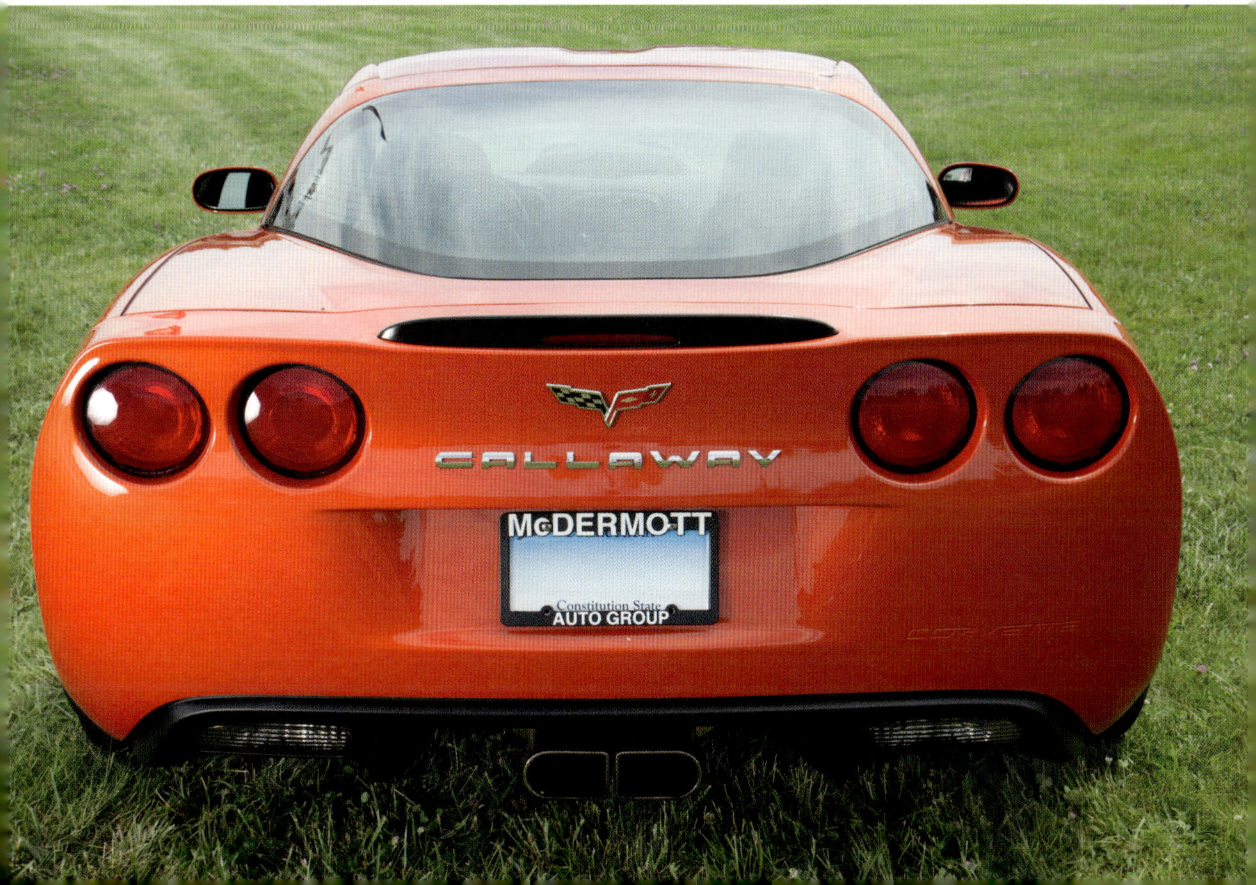
CALLAWAY
McDERMOTT
Constitution State
AUTO GROUP

Brother to Brother

1961 Convertible, Black with Red Cove
Owner: John

This 1961 Corvette was purchased by John's brother in 1971 for $900. The exterior color was silver with white coves and a red interior with a white convertible top. It had a different motor and transmission, six tail lights and a scoop hood. The tires were whitewalls with full hubcaps.

John's brother took the car apart and it was stored in two metal sheds—where it sat for three years. Finally, their father gave him an ultimatum: either put it back together or sell it. That's when John's brother began to work on it. He replaced the frozen motor with one from a 1966 Chevelle that he rebuilt. He also changed the exterior to black and the coves to red.

John bought the car from his brother in 1980. "I repainted it in 1994," he says, "and bought new Cragar wheels. I later purchased American racing wheels and had the interior redone in red and added sequential tail lights. In April of 2017 I installed a new Tremec five-speed transmission, which improves the gas mileage and provides better drivability."

John has brought his 1961 Corvette to numerous car shows and car cruises and won many trophies.

Test the Limits

2004 Commemorative Edition, LeMans Blue
Owner: John

I asked Jeffrey why he bought this 2004 LeMans Blue Commemorative Edition Corvette. He answered, "This is my first Corvette and my story is simple. I was taken with the beautiful color and the interior is very comfortable and matches perfectly."

"After going through the gears a couple of times in city conditions during the test drive, I asked if I could be a little more aggressive. The salesman replied, "It's a Corvette and the highway is a mile up on the left.'"

Taking it from zero to the "speed limit" created a permanent smile. This is Jeffrey's story and he is sticking to it!

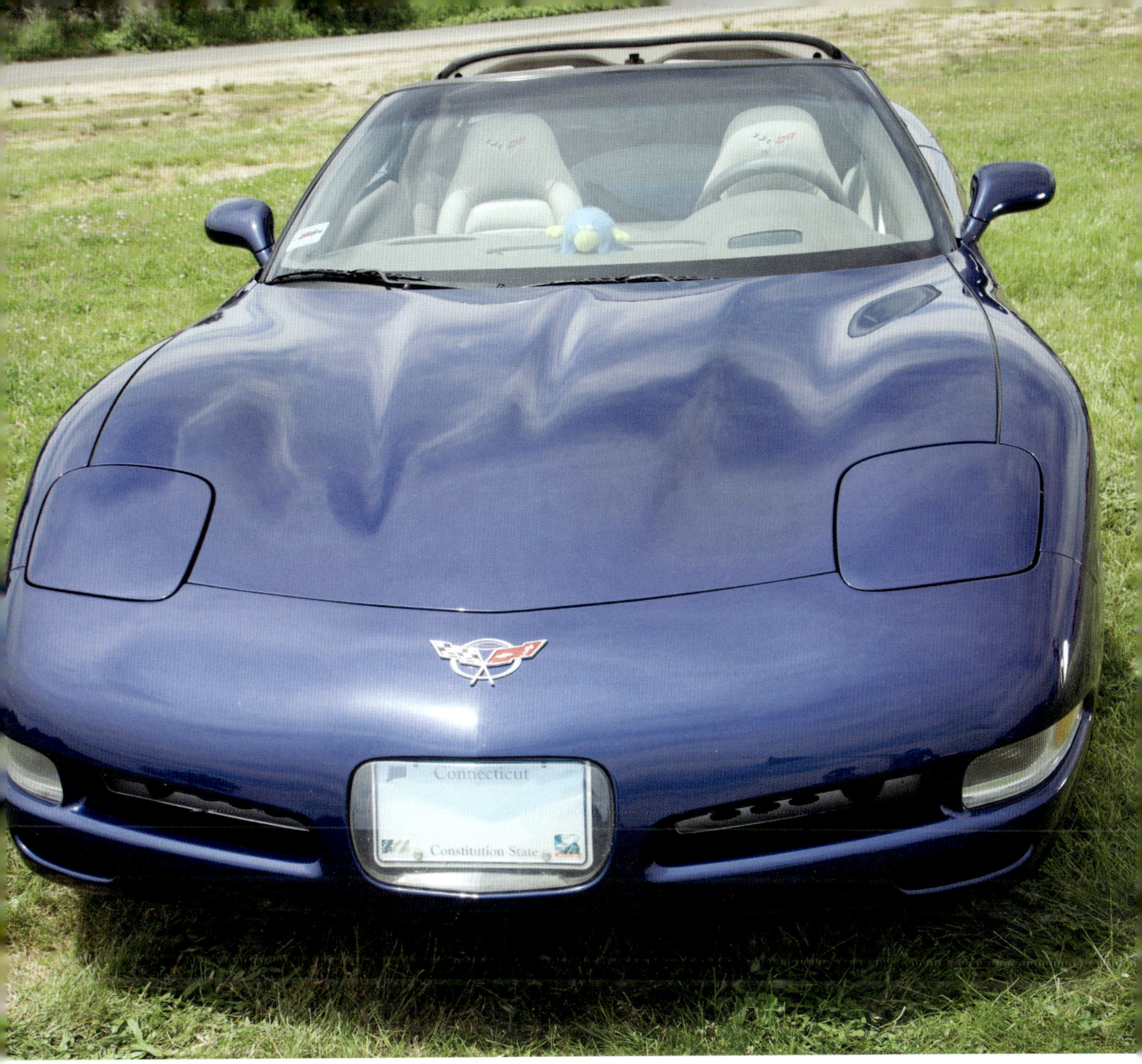
Connecticut
Constitution State
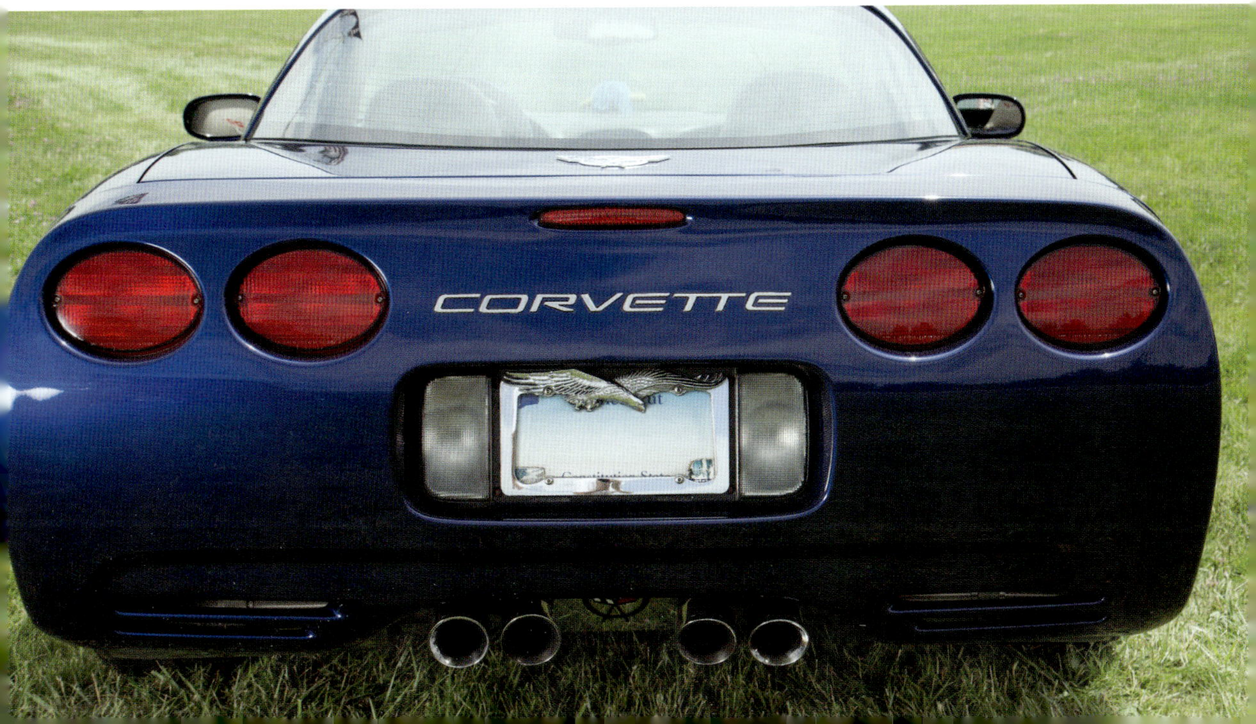
CORVETTE

On the Track

2015 Z51 C7, Shark Gray
Owner: Jed

This is Jed's first Corvette. "I went to the Ron Fellows Driving Course, a three-day racing school for Corvettes, in Pahrump, Nevada, and drove my 2015 Shark Gray Z51 C7 on the track. Not only was it a great time, but I gained a lot of information and learned many techniques about how to drive on the track," he says.

CORVETTE
www.RaceSpringMountain.com
Connecticut
SPRING MOUNTAIN

The "Corvette Guy"
2017 Grand Sport, Watkins Glen Gray
Owner: Paul

Paul's story is about a 14-year-old kid spotting a white sports car; it was a 1960 or 1961 Corvette. That was the day he became a "Corvette Guy"—one who loves Corvettes and, most of all, loves to drive Corvettes.

"I purchased my first Corvette in 1969, a 1962 Ermine White convertible," he says, "and purchased my second Corvette in the early 1970s—a 1969 Fathom Green coupe. My third was purchased in 1973, a 1973 orange coupe. I paid $5,000 for it in 1973 and sold it for the same price in 1983. I bought my fourth in 1983, a 1979 silver coupe that I had for one year. A number of years passed and I thought that was the end of my Corvette days because of family obligations."

"After our kids finished college, my wife and I downsized," Paul continues. "It was time to be a 'Corvette Guy' again. I bought my fifth Corvette in 2008, a 1999 C-5 Pewter convertible. I was certain it would be my last—but there was another part of my Corvette dream, which was to go to a Chevy dealer and order a Corvette exactly the way I wanted it. In 2017, I ordered a 2017 Watkins Glen Gray Corvette Grand Sport. It was my 70th birthday present to me, the 'Corvette Guy.'"

Childhood Memories

2005 Convertible, Victory Red
Owner: Anthony

The first time Anthony saw a Corvette was in 1970. He was sitting in the car waiting for his mother to come out of a store—and when she did, he asked her what it was. She told him that it was a Corvette. "I fell in love with Corvettes when I saw that 1969 Stingray," he recalls. "I am so thankful that I now own a 2005 Victory Red Corvette."

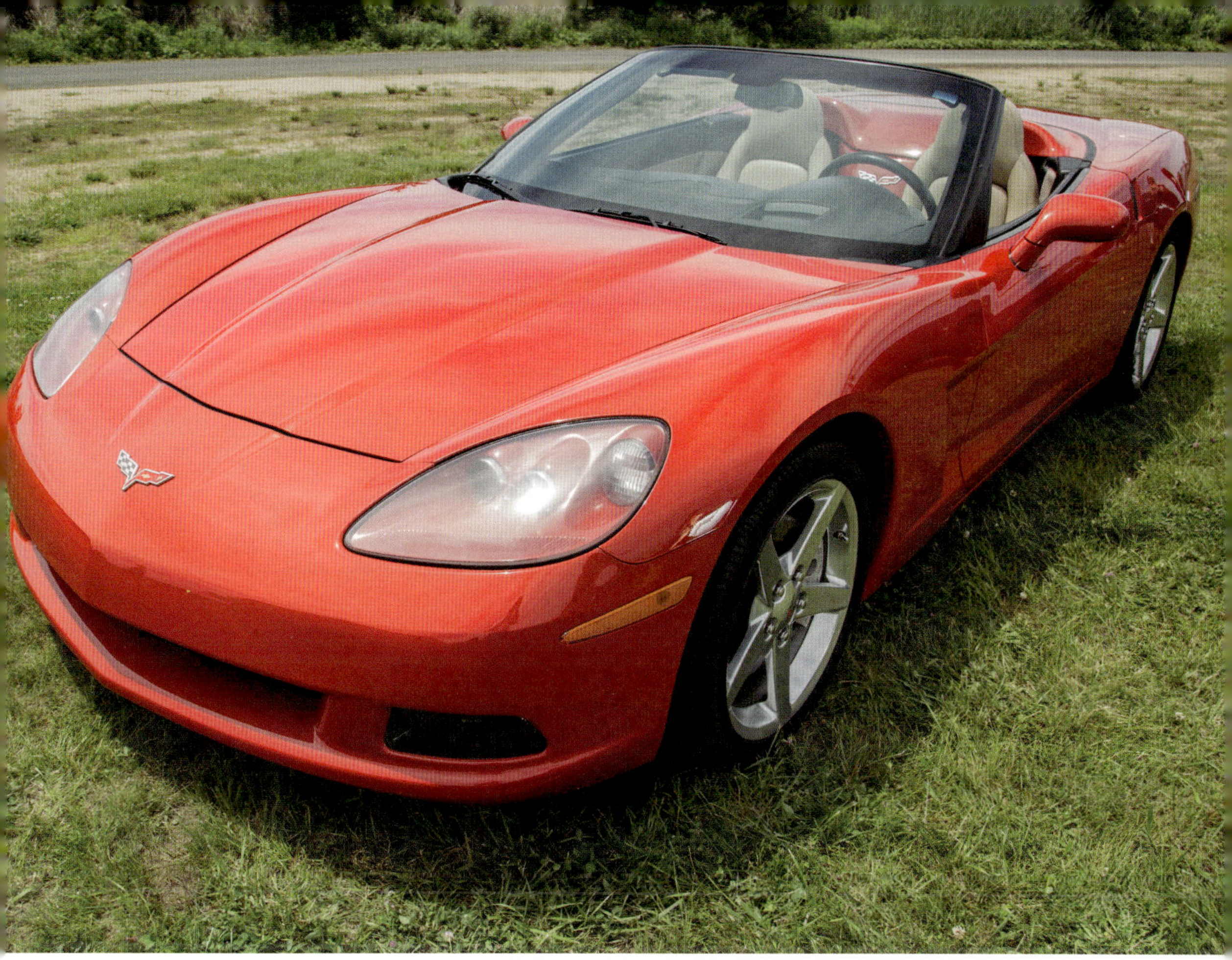

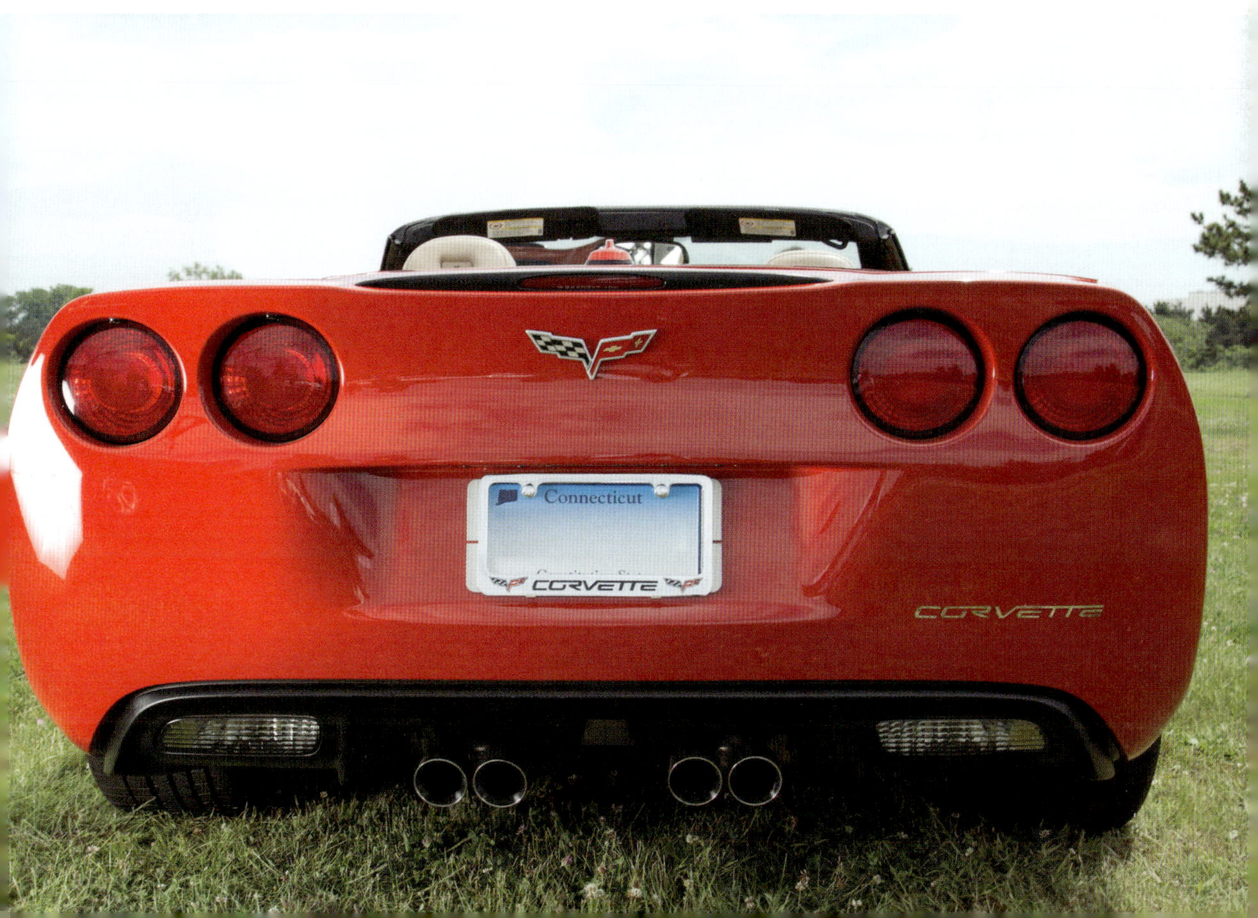
Connecticut
CORVETTE
CORVETTE

Pace Car: Indianapolis 1998

1998, Radar Blue
Owners: Tony and Rose

Tony and Rose have owned a 1978 Corvette for 37 years. They wanted to add a C5 and decided on the 1998 Radar Blue Corvette Pace Car. "It was not love at first sight," confesses Tony. "Initially we thought it was gaudy. However, we warmed up to it, bought it, and have had it for 17 years. We have had a lot of fun with it. Whenever we drive it anywhere we get comments from people about how awesome it looks. It is a great conversation piece."

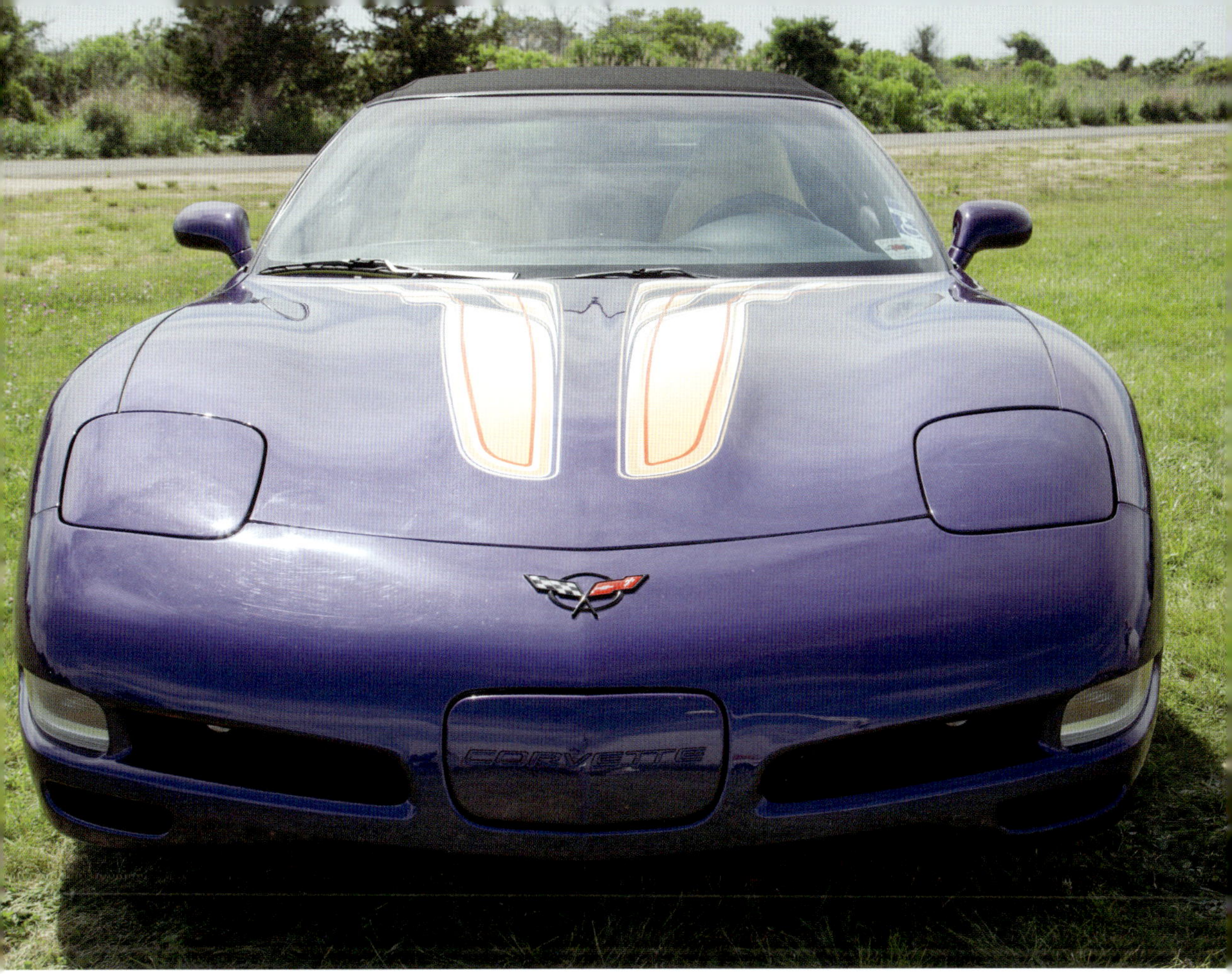
CORVETTE

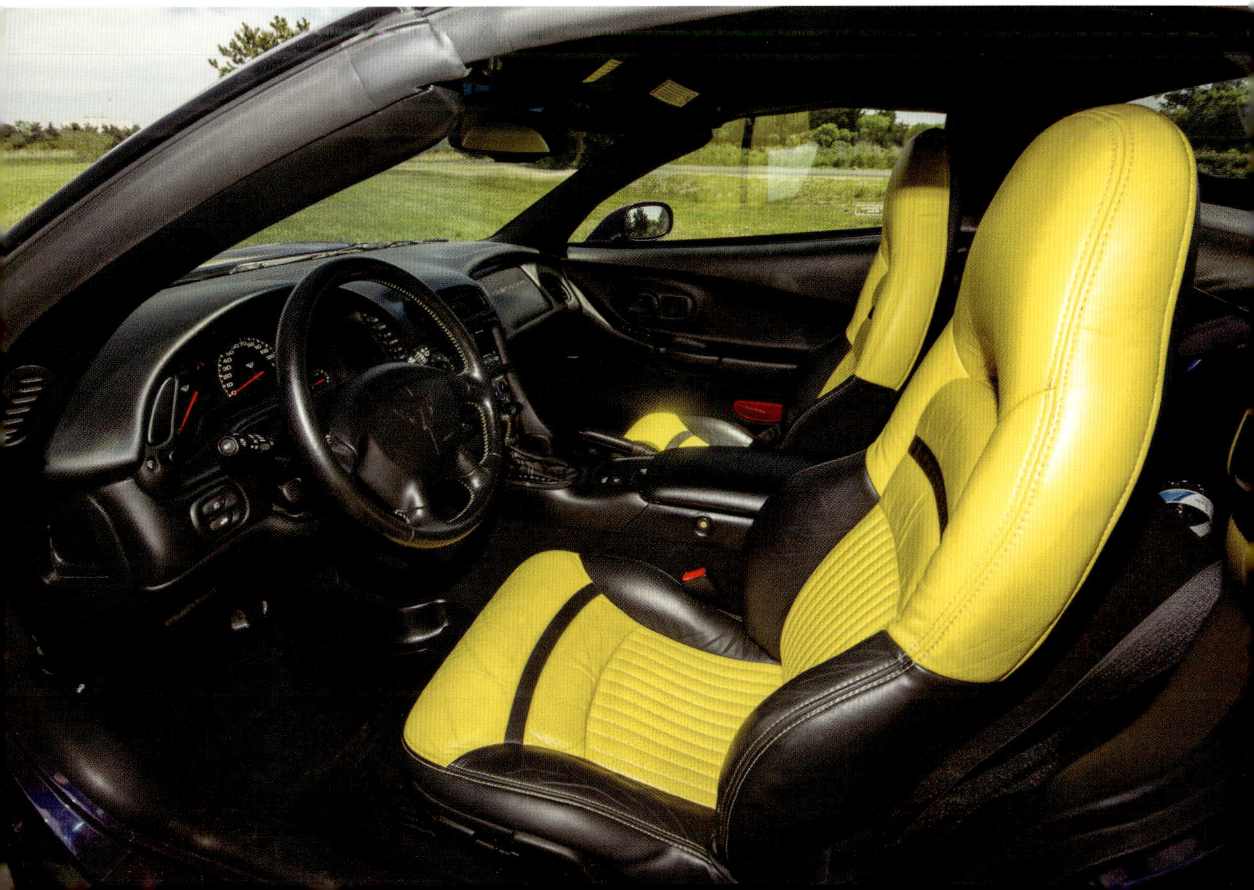

For Their 30th Anniversary

1968 Convertible, Blue
Owners: Allen and Teri

Allen and Teri discovered that they both wanted a Corvette C3 when they were kids—many years before they knew each other. It had to be a convertible with a four-speed transmission, and old enough to have chrome bumpers.

"For our 30th anniversary we decided to finally buy one and found this unrestored, very original, 1968 blue Corvette convertible with the 350HP 327 engine. We've had other fun cars, but there's something about a Corvette that makes it special. We drive the car weekly in-season, and no matter how much work we do to it, we plan for it to always be a driver," says Allen.

CONNECTICUT
Classic Vehicle

CORVETTE
CONNECTICUT
Classic Vehicle

In Honor of a Friend

1962 Fuelie, Honduras Maroon
Owner: Steve

Steve is one of three people with two Corvettes in this book. His story about his 1966 Corvette appeared on page 60.

While Steve was restoring his 1966 Corvette, a friend was also restoring a Honduras Maroon 1962 Corvette Fuelie. Sadly, his friend passed away suddenly from a heart attack in the winter of 2008. At that point, the restoration on the 1962 was only half complete. Steve approached the executor of his friend's estate and offered to purchase the car in order to complete the restoration to NCRS specifications. The restoration was finished in 2017. Today, Steve and his brother Mike show the car at local cruise nights and car shows to honor their friend's memory.

Two for One

2017 Grand Sport, Long Beach Red
Owner: Richard

Richard's first Corvette was a 2001 Dark Bowling Green Metallic convertible. In 2009, he purchased a 2009 Crystal Red coupe and maintained both cars until he started looking at the C7.

"I gave my brother the 2001 in order to make room in the garage for a 2016 Stingray," says Richard. "Realizing that two cars take time to maintain, I then traded both cars in for a 2017 Long Beach Red Grand Sport. I convinced my wife that this was a good idea by telling her I was downsizing and getting money back." He adds, "As I get older and my time gets shorter, I need a vehicle that gets me places quicker. The Corvette does a great job of that."

CORVETTE
CLUB CORVETTE
Connecticut
Constitution State
OF CONNECTICUT

First Dibs

1963 Convertible, Black
Owner: Martin

Martin bought this 1963 black Corvette in December of 2011. It had been refurbished in 1988, but the previous owner rarely used it and only drove it about four times.

"In 2003, I asked the owner if he wanted to sell this Corvette," says Martin, "but it was not for sale at that time. Whenever I saw him over the next eight years, I asked about the car. He eventually told me that I would get first dibs when the time came for him to part with it."

"In 2011, he called to say that the car was for sale. Our arrangement was that I would only have to paint his truck, plus give him a couple thousand dollars. I think I made out pretty good with this deal," he laughs.

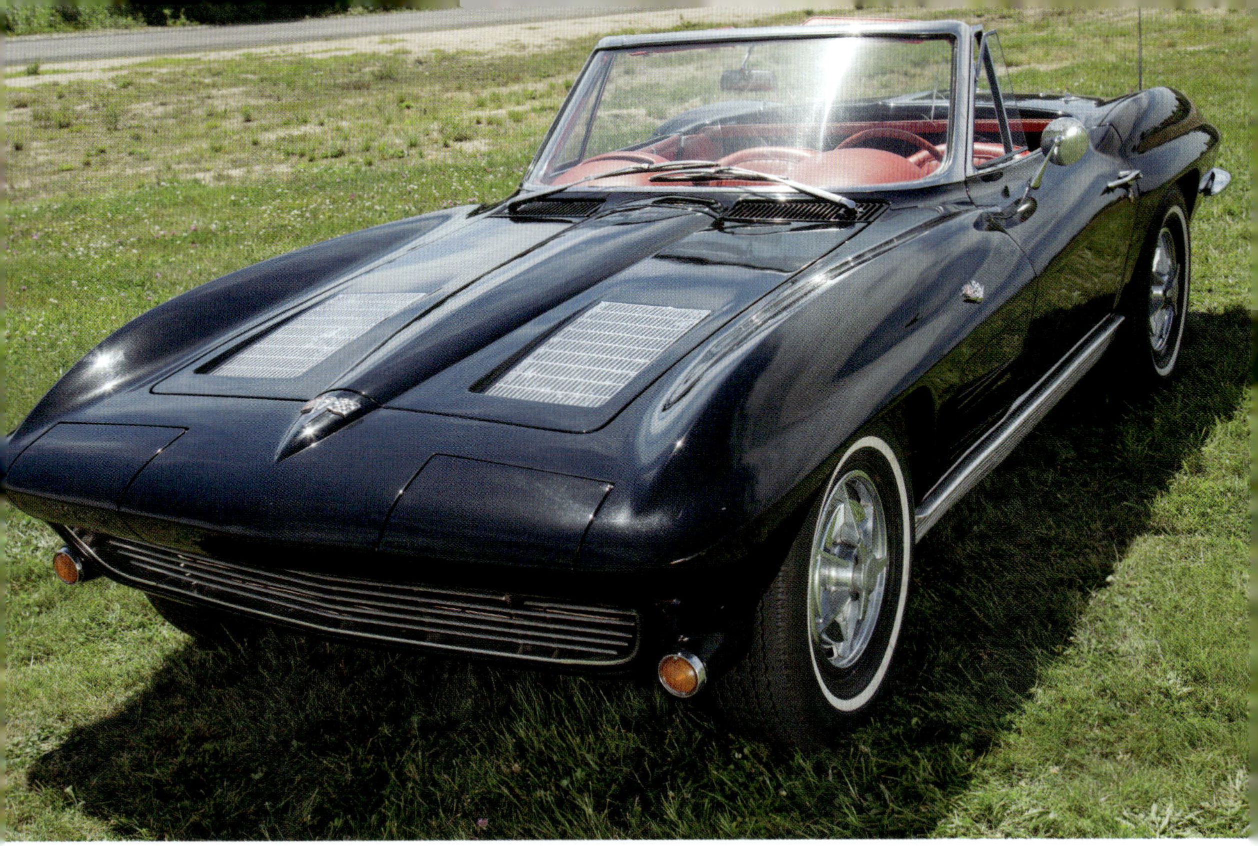

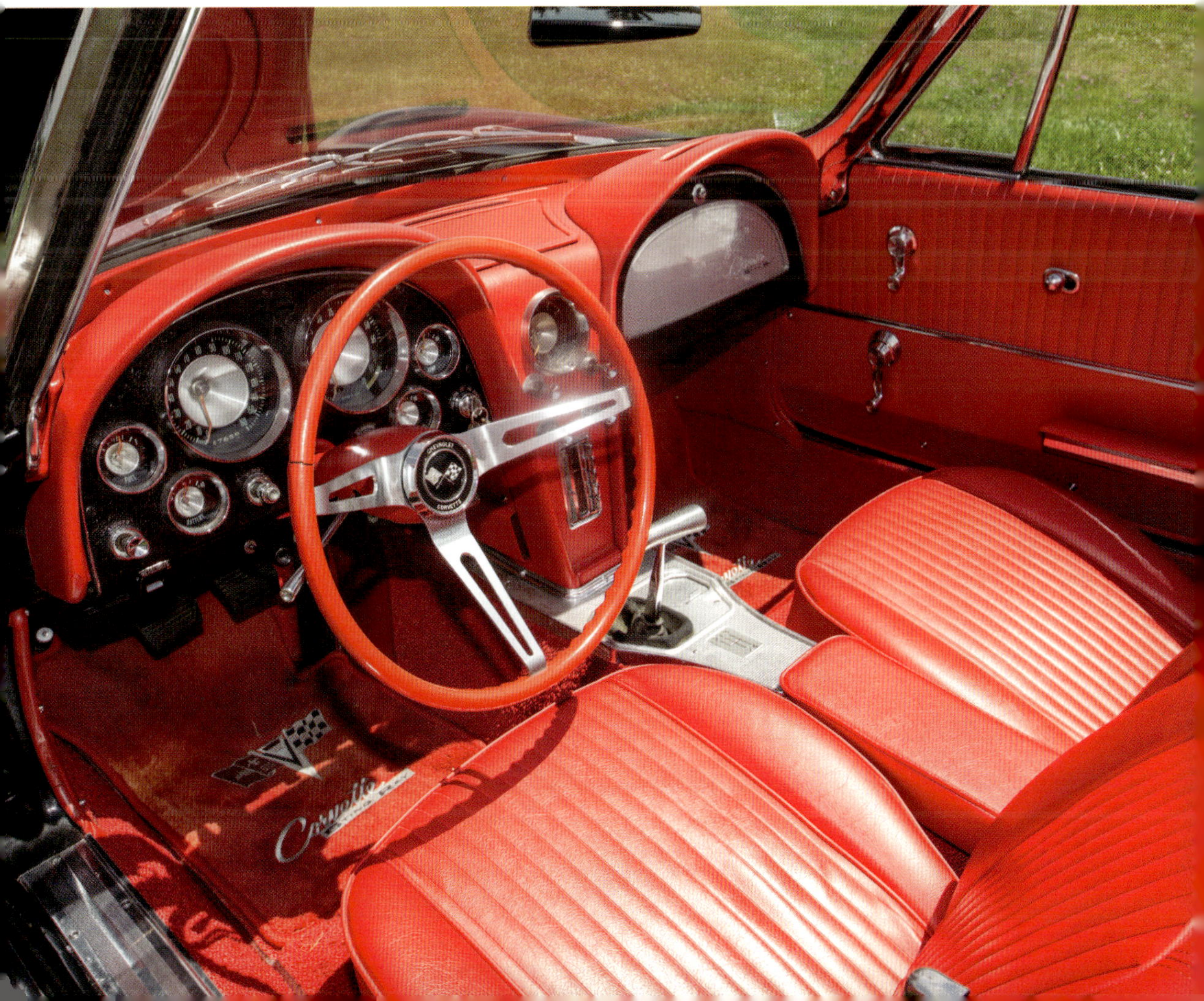
Corvette

A Rare Engine
1996, Competition Yellow
Owner: Paul

Paul's Corvette is a 1996 in Competition Yellow with the LT4 engine option. It has a manual six-speed transmission along with a heavy-duty Dana 44 3.42 rear end.

Very few Corvettes were built by GM with an engine option that was offered for only one year. This LT4 engine is one of them, which makes it a rare car. It is documented by the Corvette Museum as the 97th car out of 103 that were made.

Paul's previous Corvettes were a 1963 Roadster, a 1966 coupe, a 1968 convertible, a 1969 convertible, and a 1974 coupe.

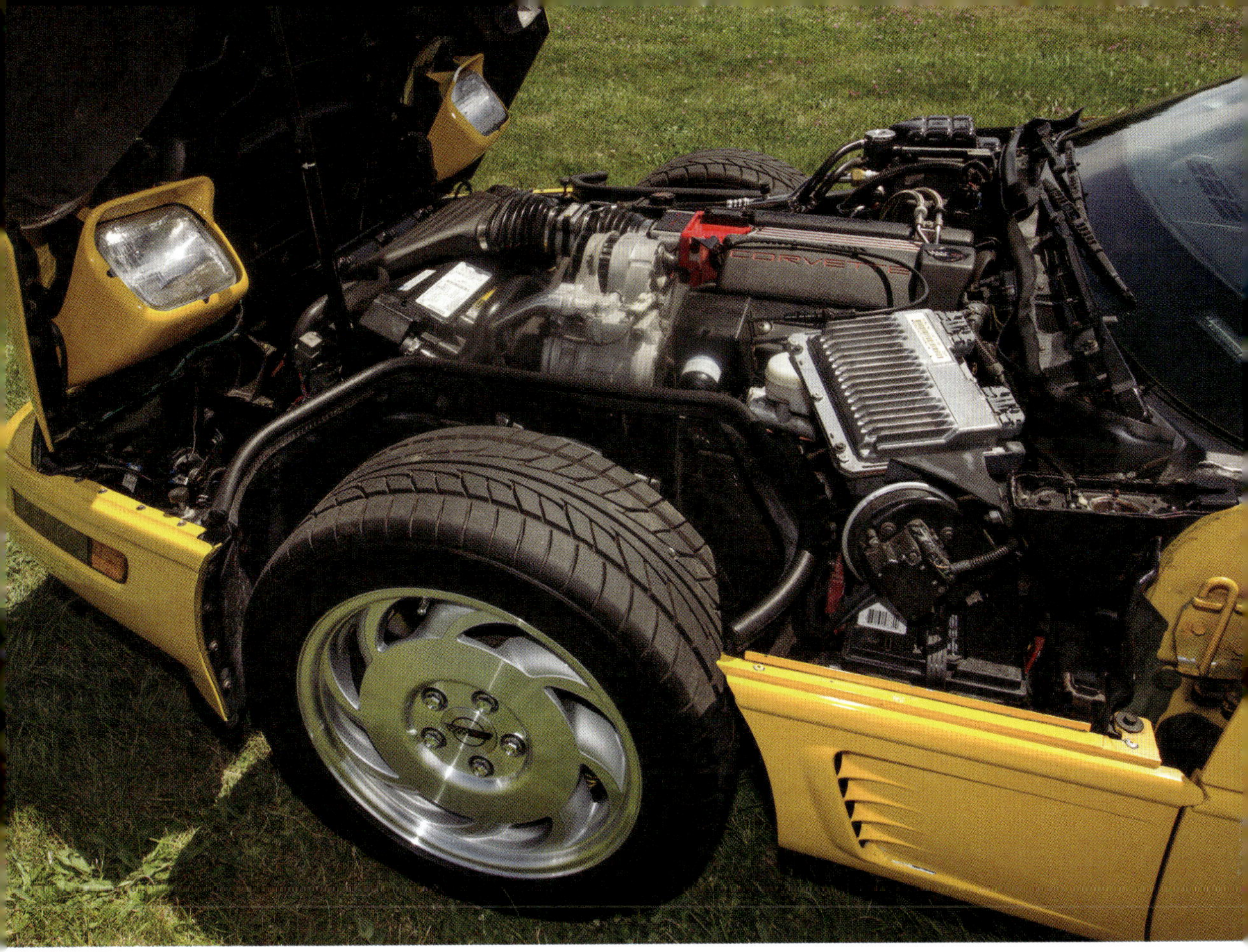

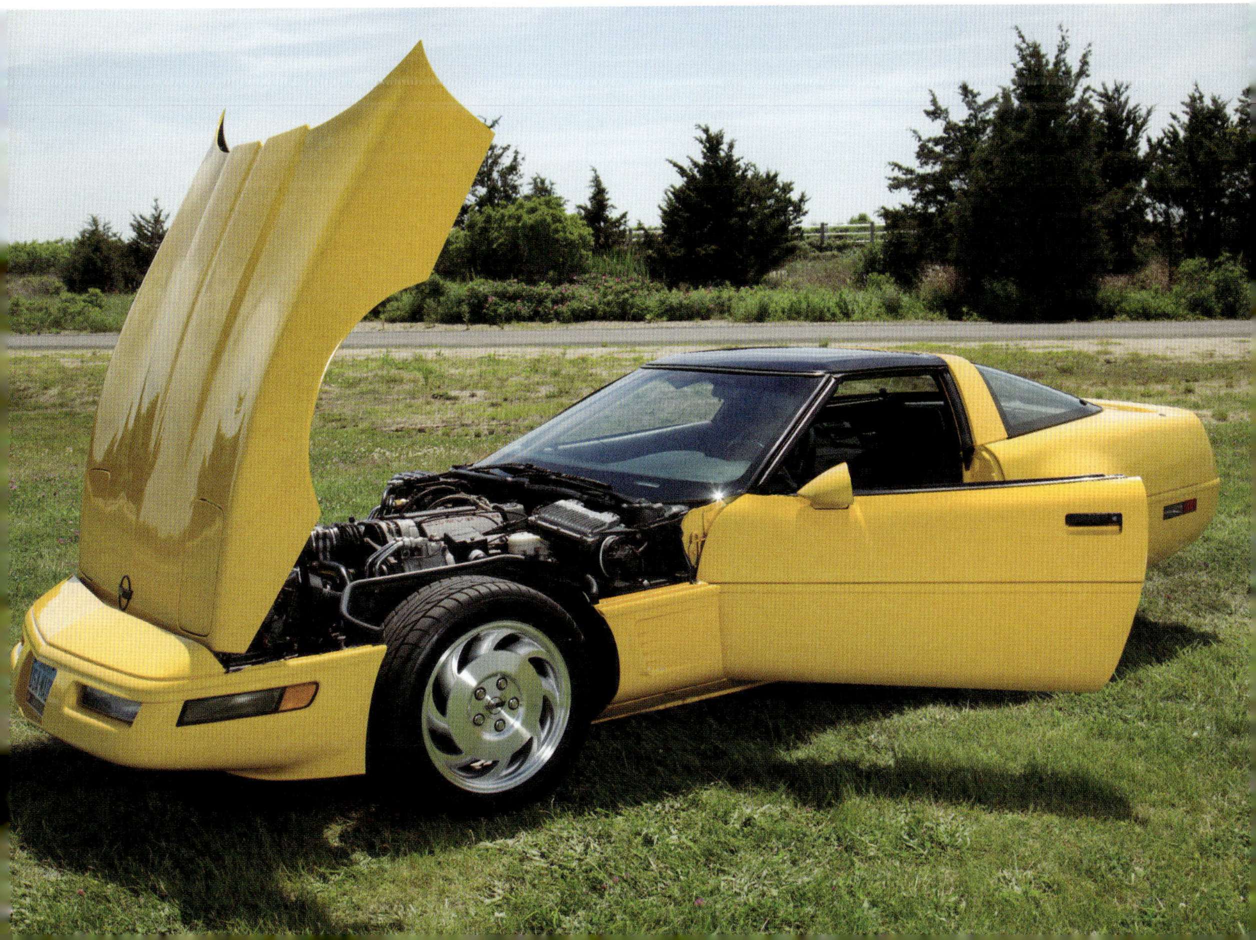

To Remember a Friend

1960 Convertible, Red with White Cove
Owner: Terry

Before finding this red convertible, Terry had been looking for a 1960 Corvette for three years because of an adventure in 1969.

"In 1969, a fellow Army trainee and I took a journey across America in a 1960 Corvette to Seattle, Washington, which was our staging area for Vietnam. The memories we created on this trip lasted a lifetime," he says. "The Corvette Connection played a key role in my acquiring this 1960 red Corvette with the white cove. Every time I get behind the steering wheel, I remember my friend who did not come home from the war and the great times we had on our journey."

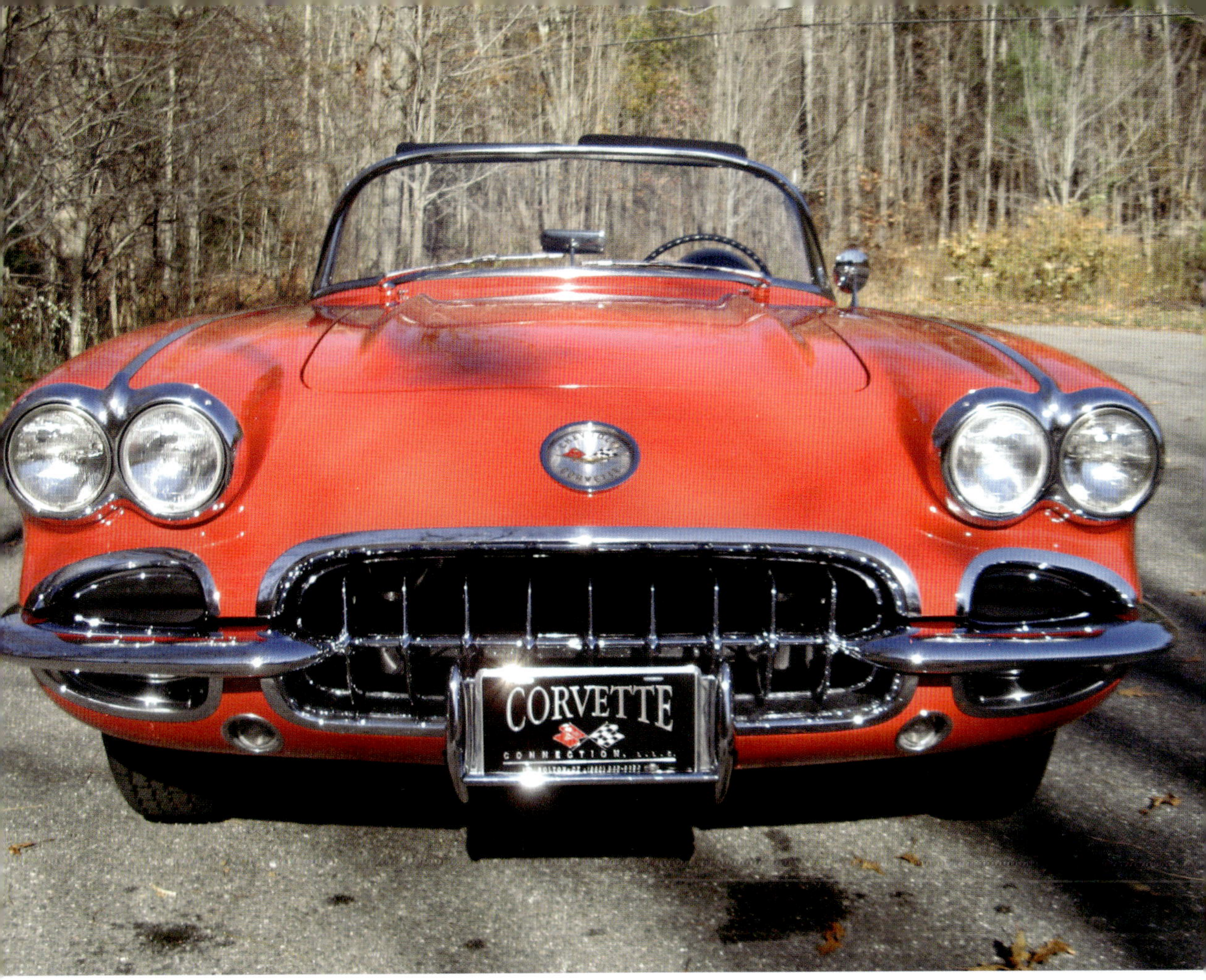
CORVETTE

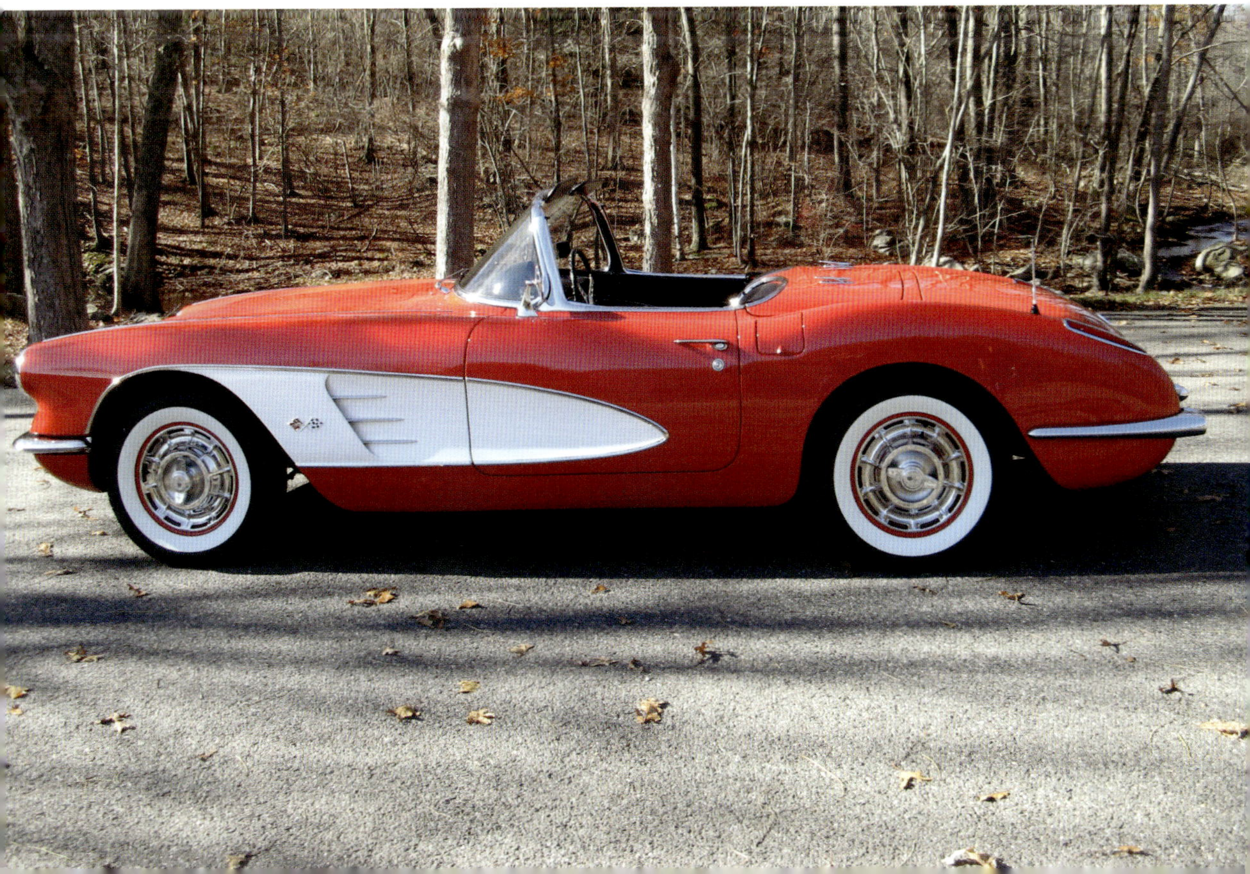

In Honor of Warriors

2004, Magnetic Red Metallic II

Owner: Sharon

This car holds a special place in Sharon's heart. She has been a Corvette junkie since she was a kid; as an adult, she made her dreams come true and became an owner.

"I already owned a 2015 Stingray when I saw this beautiful 2004 Magnetic Red Metallic II Corvette at a car show. There was a 'for sale' sign in the window and I could not get this car off my mind the rest of the day. When I went back to get the contact information, it was *gone!*" says Sharon. "I researched it online and found it in the neighboring town. The car was originally from Dallas, where a man purchased it for his wife who was dying of cancer. It had been on her bucket list to own and drive a Corvette. After she passed away, he couldn't bring himself to drive it or part with it. He eventually sold it to a dealer and hoped that it would go to a good home."

"Our Corvettes speak to us," says Sharon. "We make an emotional connection to these special automobiles. As a tribute to breast cancer warriors everywhere, I drive my Corvette in your honor."

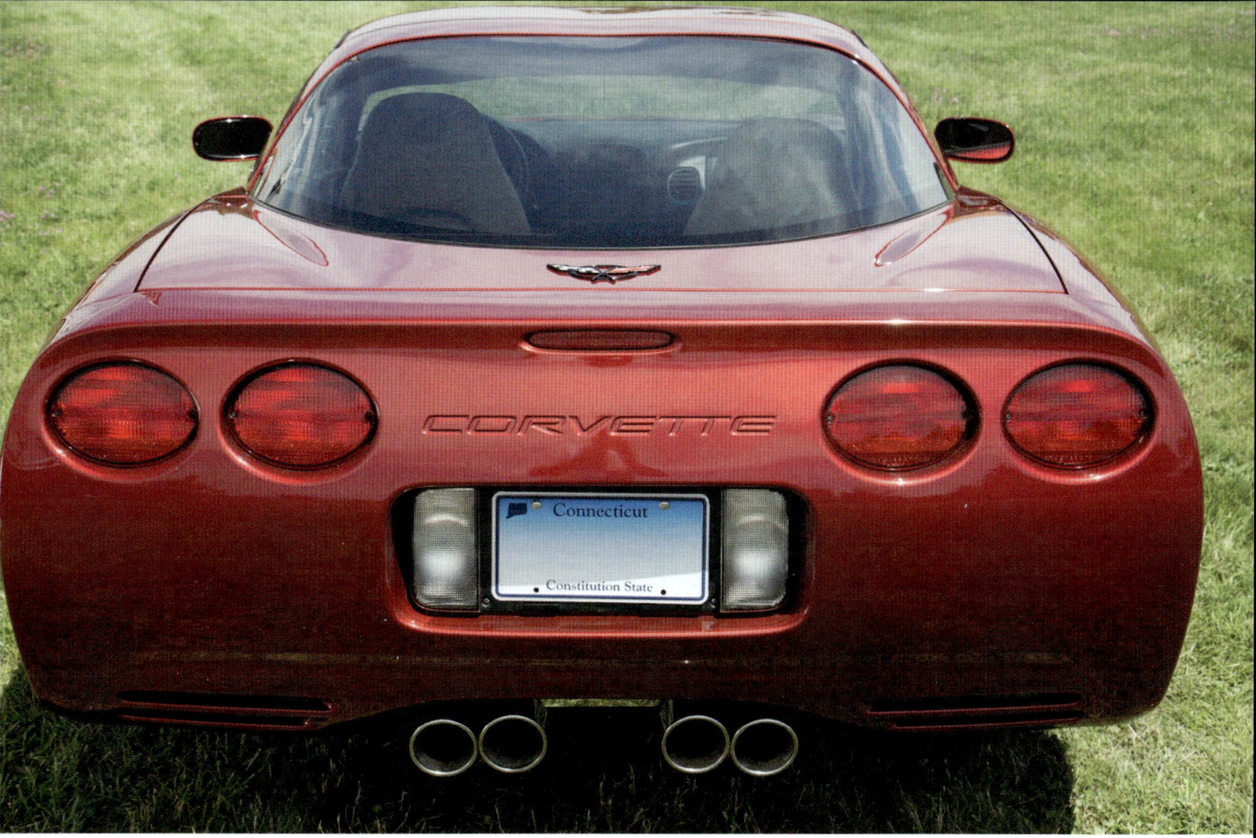
CORVETTE
Connecticut
Constitution State

Connecticut
Constitution State

Pace Car: Indianapolis 1995

1995, Purple and White
Owner: Donald

When Donald first saw the 1995 pace car, he was not fond of the purple and white color combination. However, a friend had one and the car grew on him.

"In the early 2000s, I began to look for one," he says. "The first one I found slipped through my fingers as we negotiated a price. When a second opportunity arose, my wife told me to bend a little more on pricing. When I drove it home her response was, 'Could you have found a gaudier car?' Sometimes you can't win."

WASHER FLUID
ONLY

CORVETTE
INDY 500
CONNECTICUT
Classic Vehicle
INDIANAPOLIS MOTOR SPEEDWAY

All Over Again

2004 Convertible, White

Owner: Rich

Several of Rich's friends had Corvettes while in their teens and twenties, but not Rich—because it was not practical. Over thirty years later, long after his friends had sold their original Corvettes, Rich finally bought his first: a white 2004 Corvette convertible. "Many of my old friends have purchased new ones," he says. "We all have at least one and we are enjoying them all over again."

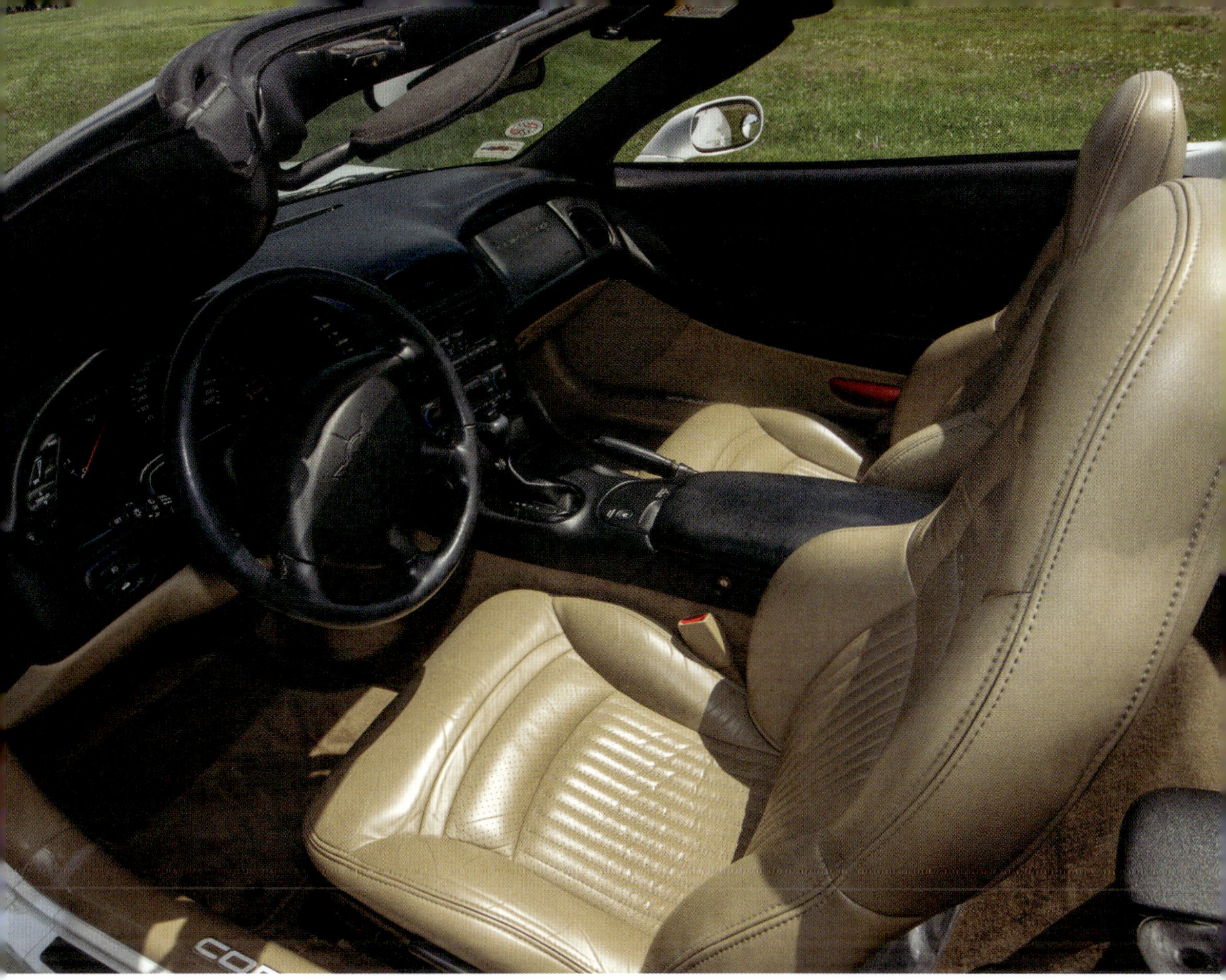

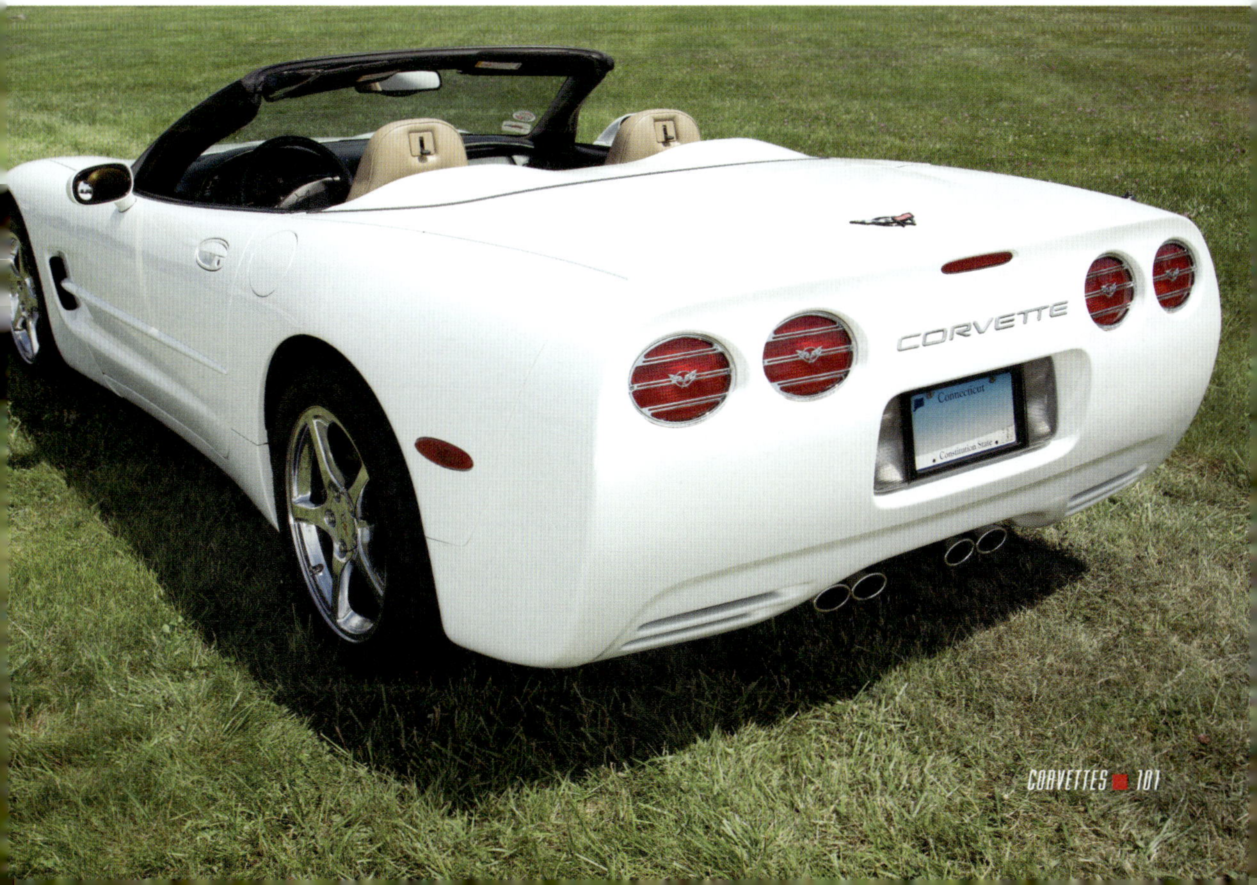
CORVETTE
Connecticut
Constitution State

A Real Looker

1962 Stingray, Fawn Beige
Owner: Ed

Ed purchased this Fawn Beige 1962 Corvette with Fawn Beige interior in 2006.

"I love this four-speed 327 with 340 horse-power. I get to enjoy it with a hardtop as well as a white soft top in nice weather," says Ed.

"There were some problems with it when I first bought it. It didn't run. But most of the kinks are out and I love the looks I get from other drivers and people on the street when I pass by in this classic Corvette Stingray."

On the Spot

1996, Torch Red
Owner: Stephen

Stephen has owned six Corvettes over the years. He lost his first one, a 1970 LT-1 coupe, in an accident. His next two did not take its place; he sold them and waited five years before he bought a new 1981 Corvette, which he had until the 1984 models came out. "The '84 was an interesting car, but the ride was not great," he says. "I saw and loved a Torch Red 1994 with red interior—but with two kids in college, the timing was not right."

"In the spring of 1998, I began searching the internet and found this Torch Red 1996 Corvette with a red interior in Ohio. I inquired about it, but I never thought I would get to Ohio. When I went to Buffalo on business, I found that the car was still available. I drove from Buffalo to Ohio and bought it on the spot."

"I have done a lot to it, as I enjoy shows, and it is a consistent winner. I also take occasional trips to the drag strip. I changed the rear end to a Dana 44 with 3.54 gears that came out of a 1992 Corvette," says Stephen.

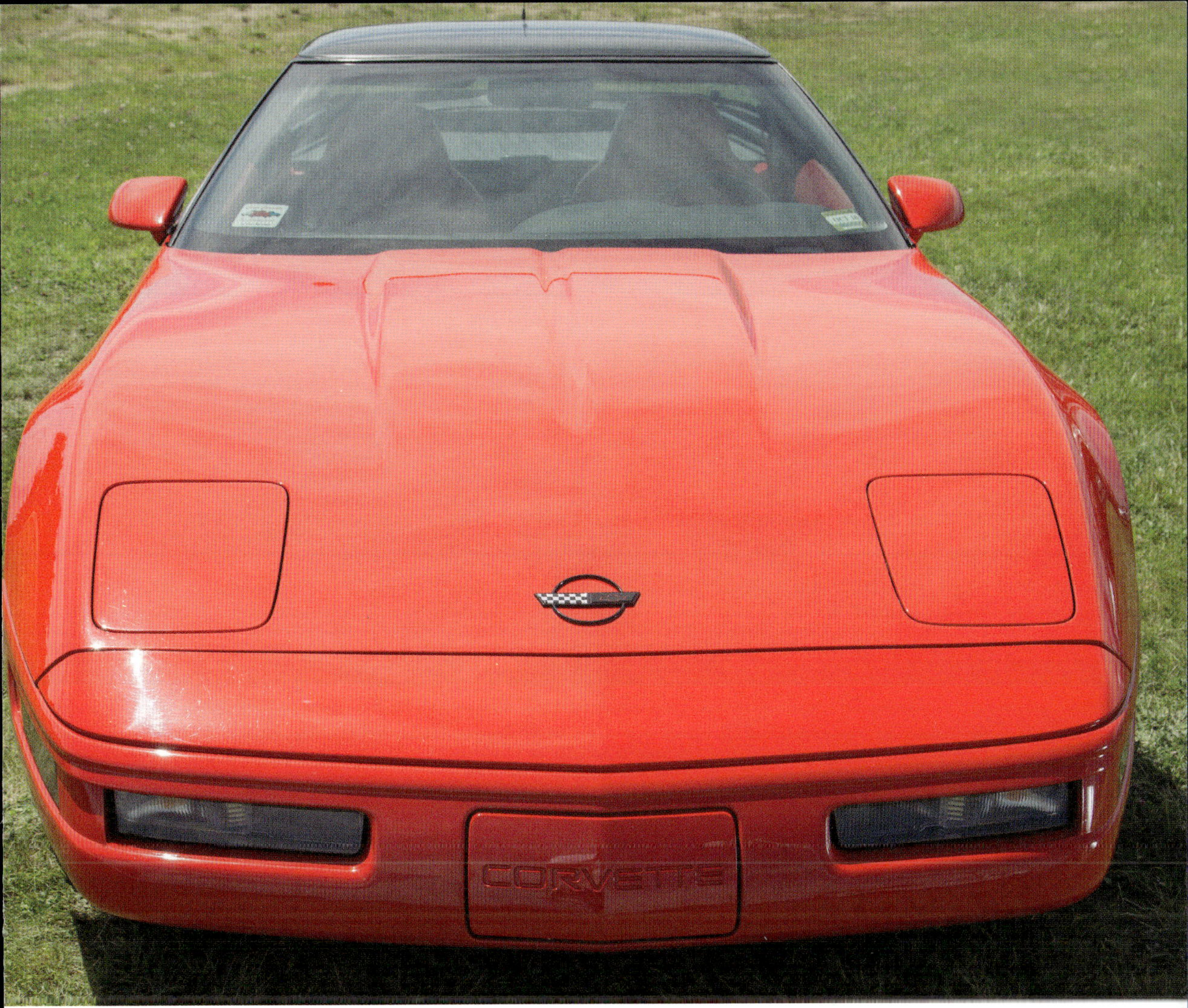
CORVETTE

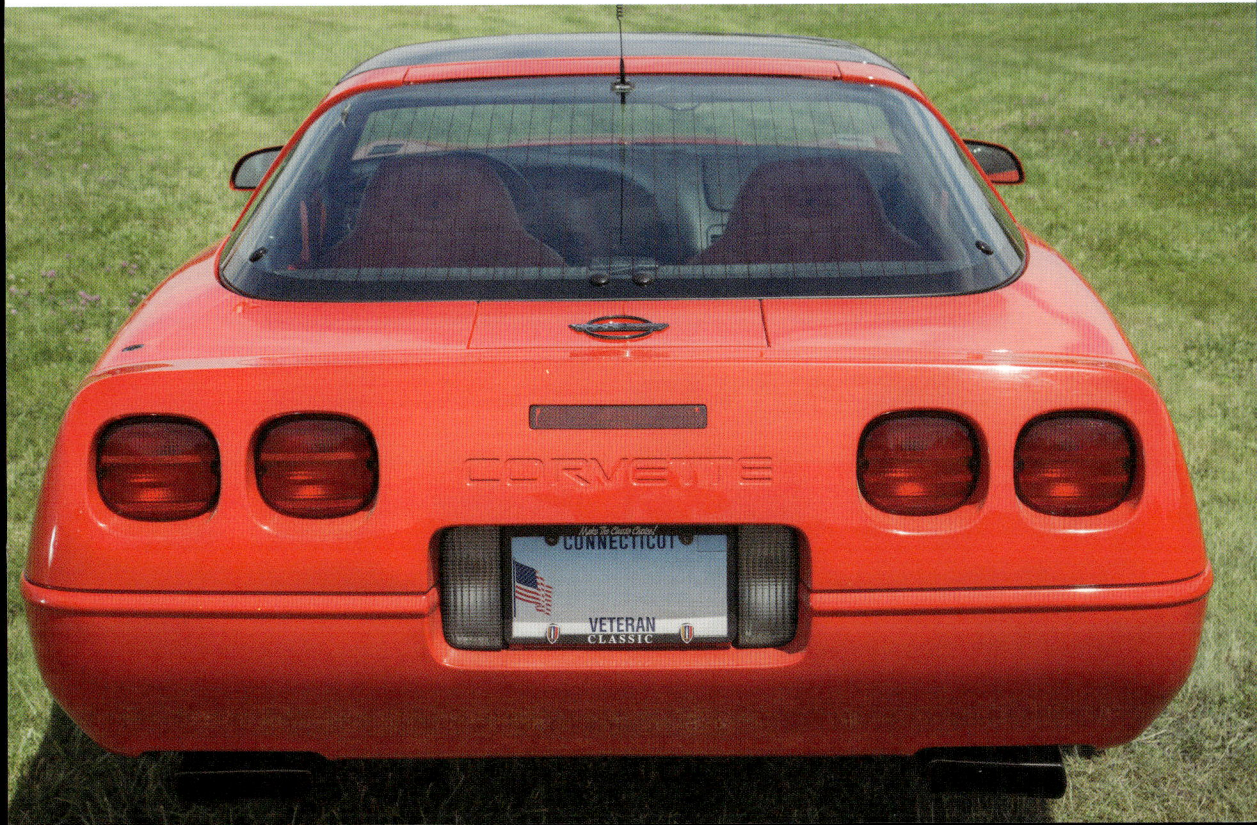
CORVETTE
CONNECTICUT
VETERAN
CLASSIC

![Silver 1979 Corvette parked on grass]

A Sweet Surprise

1979, Silver
Owner: Janis

Janis has a special story about her Corvette. "While visiting several Chevrolet dealerships in search of a new SUV, every time I looked for my husband he was sitting in the Corvette featured in the showroom. I decided to buy him a used one for Christmas, so the two of us could enjoy it together. I contacted a broker in the south, gave him my budget, and sent him off on a search."

When the broker took in a 1979 silver Corvette within her budget, he said she should come down to check it out. "I bought a one-way ticket to Chattanooga and closed the deal," says Janis. "Not knowing how mechanically sound the car was, I drove over 500 miles without turning off the engine, for fear it might not restart. After stopping for the night in Allentown, Pennsylvania, the car started up the next morning, and I finished my journey home."

Janis washed and waxed the car and put a big red bow on it—much to the envy of the husbands in the neighborhood! She says, "My husband came home later that day and was stunned!"

Connecticut
Constitution State

corvette
Connecticut
CLUB CORVETTE
OF CONNECTICUT

Part of Growing Up

1962 Convertible, Maroon

Owner: Tom

This is not just a car to Tom; it was a part of growing up. He recalls, "When I got this 1962 maroon Corvette, it was about a year old with only 8,000 miles on it. The original owner bought it and then got married. He and his new wife drove the car to Florida and when they returned, they traded it in to a dealer where my father worked. I had my eye on an Austin-Healey; my father suggested that I look at the Corvette. That was it."

"The car has never been repainted, re-chromed, or rebuilt. The engine is original as is the interior. Only the soft top has been replaced. The car has always been garaged, even when I was in college. I have always taken care of the car myself, with the exception of some mechanical work done by the Corvette Connection," says Tom. "I have owned my Corvette for 54 years and still enjoy taking it out for a ride."

A Continuous Joy

1961 Convertible, White
Owner: Craig

Craig's 1961 Corvette was something he had dreamed of owning since he first saw it in a national car magazine late in 1960.

"My enthusiasm was fueled even more by watching Tod and Buz see the USA in Tod's brand new 1961 Corvette every week via the *Route 66* television program," he says.

"This Corvette has been a continuous joy to own and drive. It has needed only routine maintenance and service and has never left me stranded anywhere. I get waves or thumbs-up signs from other motorists or pedestrians every time I drive it. I know of no other car that is more fun to see or be seen in."

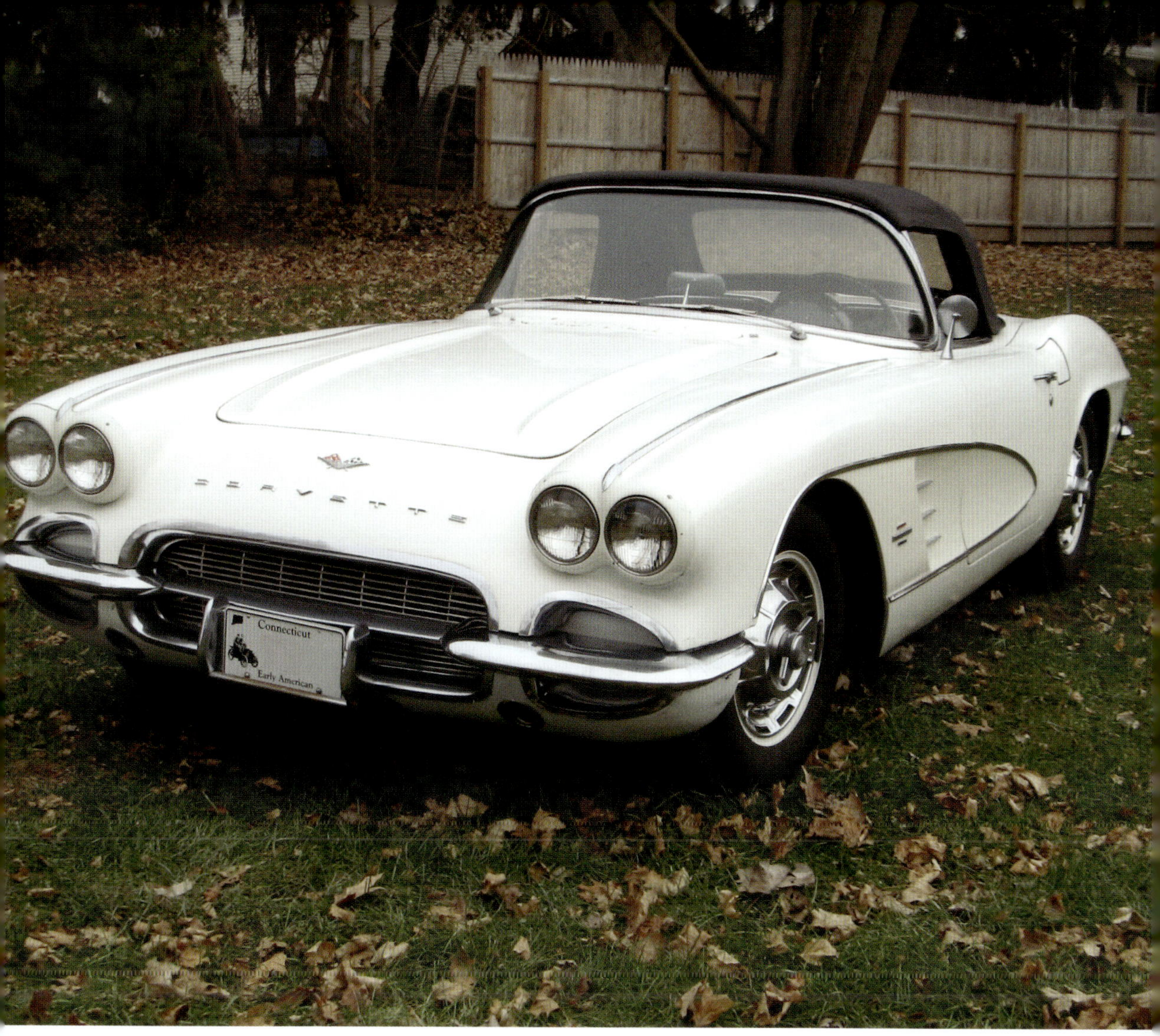
CORVETTE
Connecticut
Early American

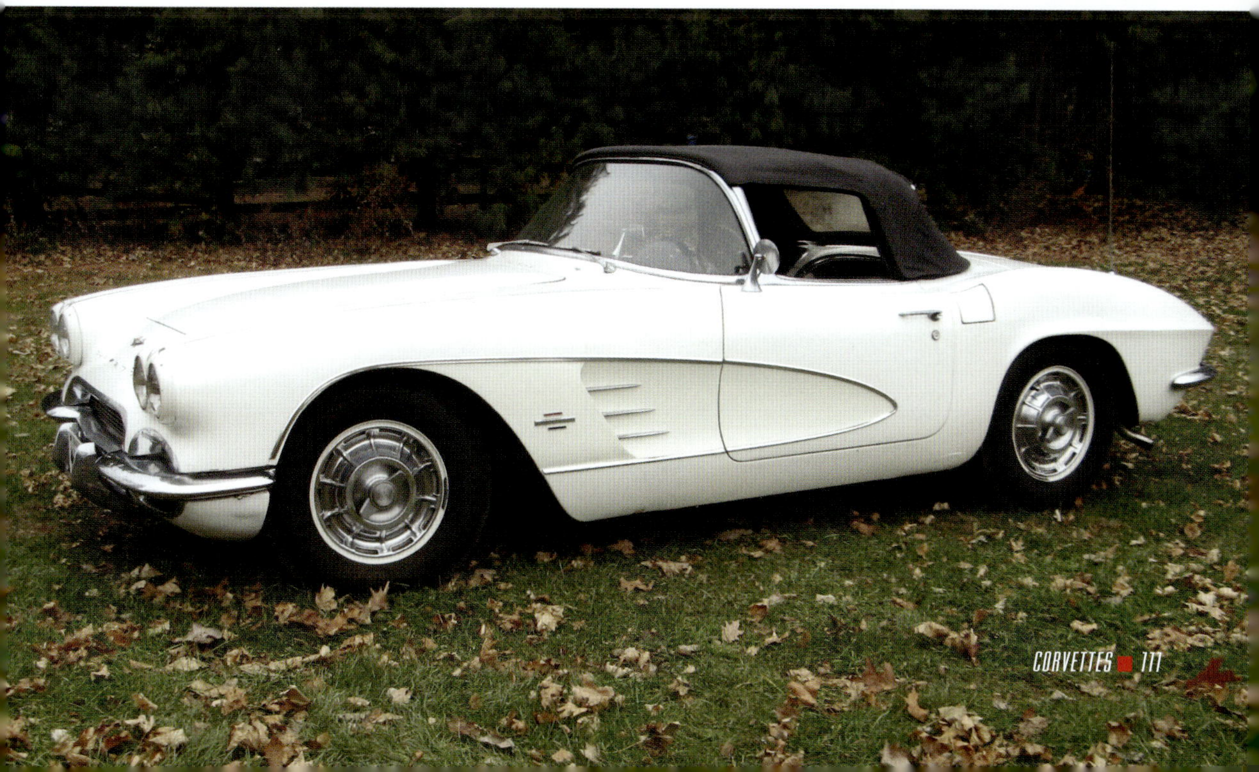

Wait for the Perfect One

1999 Convertible, Navy Blue
Owner: Anthony

Anthony waited 59 years to purchase his first Corvette. "I did a lot of research and it took over one year to find the perfect color and options for me," he says. "After waiting for so many years to own a Corvette, I was willing to wait for that certain, special car. This 1999 Navy Blue Corvette convertible is that special car!"

Head Turner

1967 Convertible, Sunfire Yellow
Owner: Ray

Ray is the proud owner of more than one Corvette; his 1965 Ermine White is featured later in this book. This is his newer model: a 1967 Sunfire Yellow with black interior.

"I love this car, with its 327-300 HP, automatic shift, power steering, and power brakes," he says. "The side pipes, bolt-on wheels, and the redline tires give this Corvette its impressive appearance. The newer models may get some attention, but people stop, turn, and look when I drive by in this vehicle."

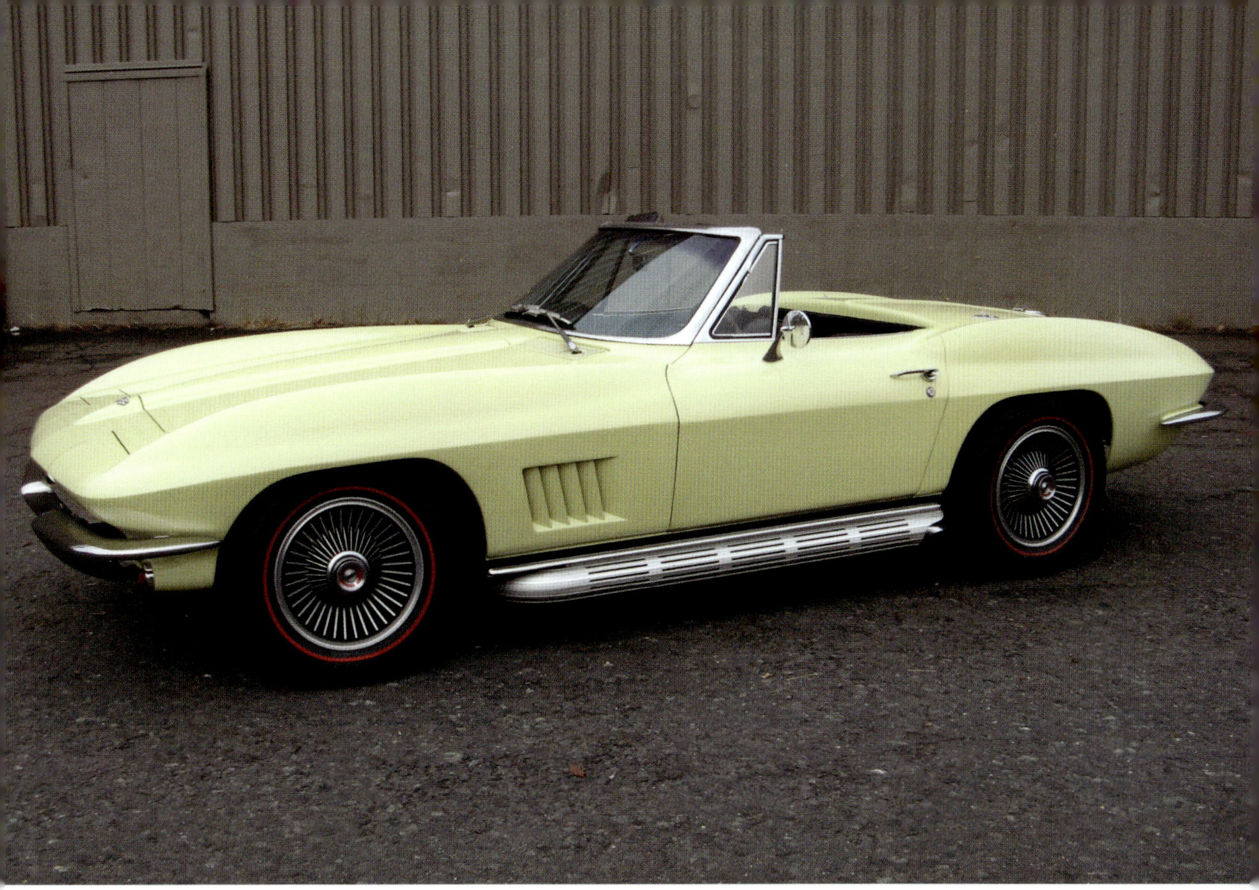

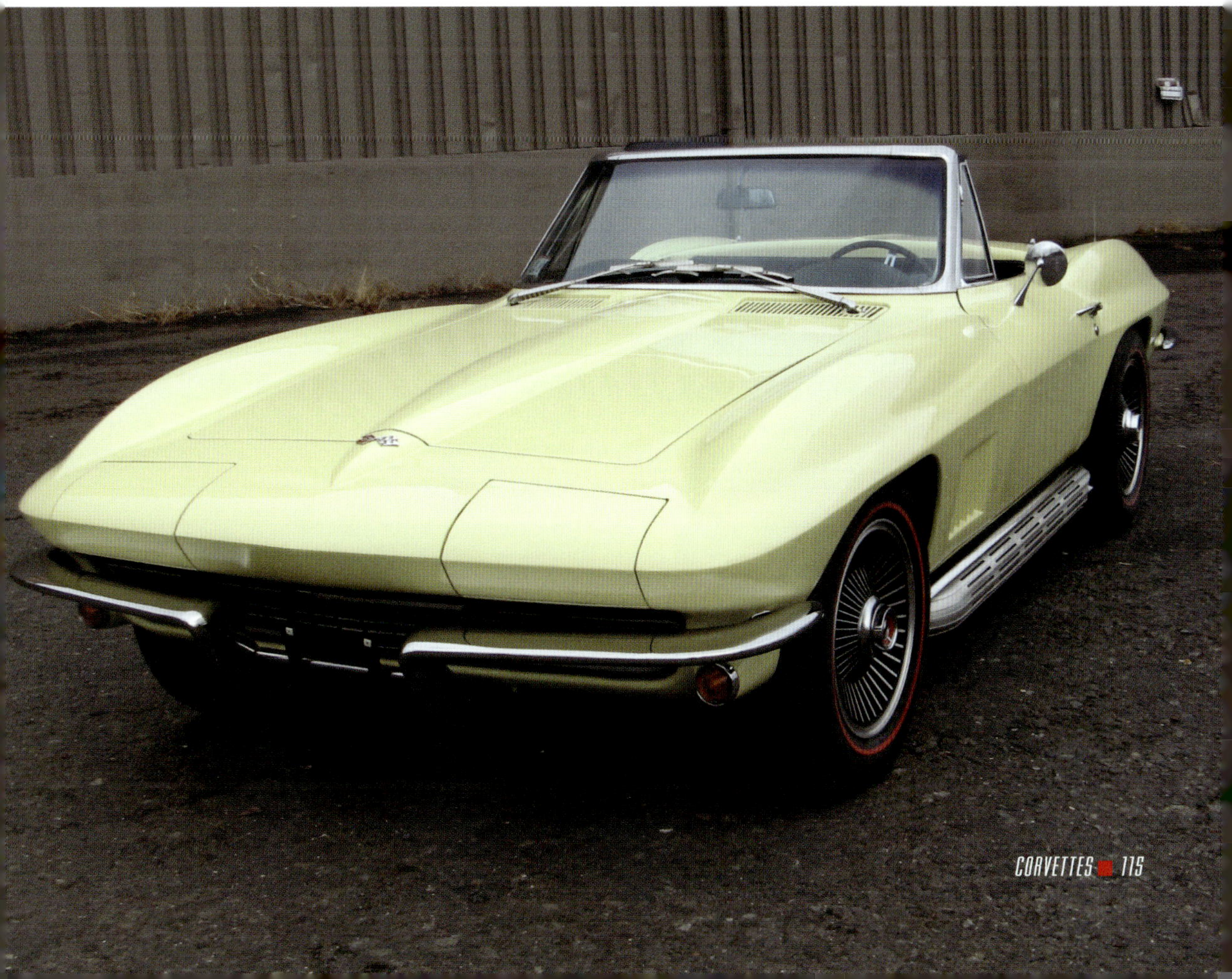

![Red 2006 Corvette convertible parked on grass]

A Wish Fulfilled

2006 Convertible, Magnetic Red
Owners: Frank and Sheryl

When Frank and Sheryl were dating, he showed up at her house in his friend's 1963 split-window Riverside Red Corvette coupe. Their love for Corvettes began that day, but they never thought they would own one.

However, in 2015, 52 years later, they bought their first Corvette. Ten months later, they bought their second—this beautiful 2006 Magnetic Red Corvette convertible.

"These amazing cars have enhanced our retirement, along with the camaraderie of other Corvette owners," says Frank.

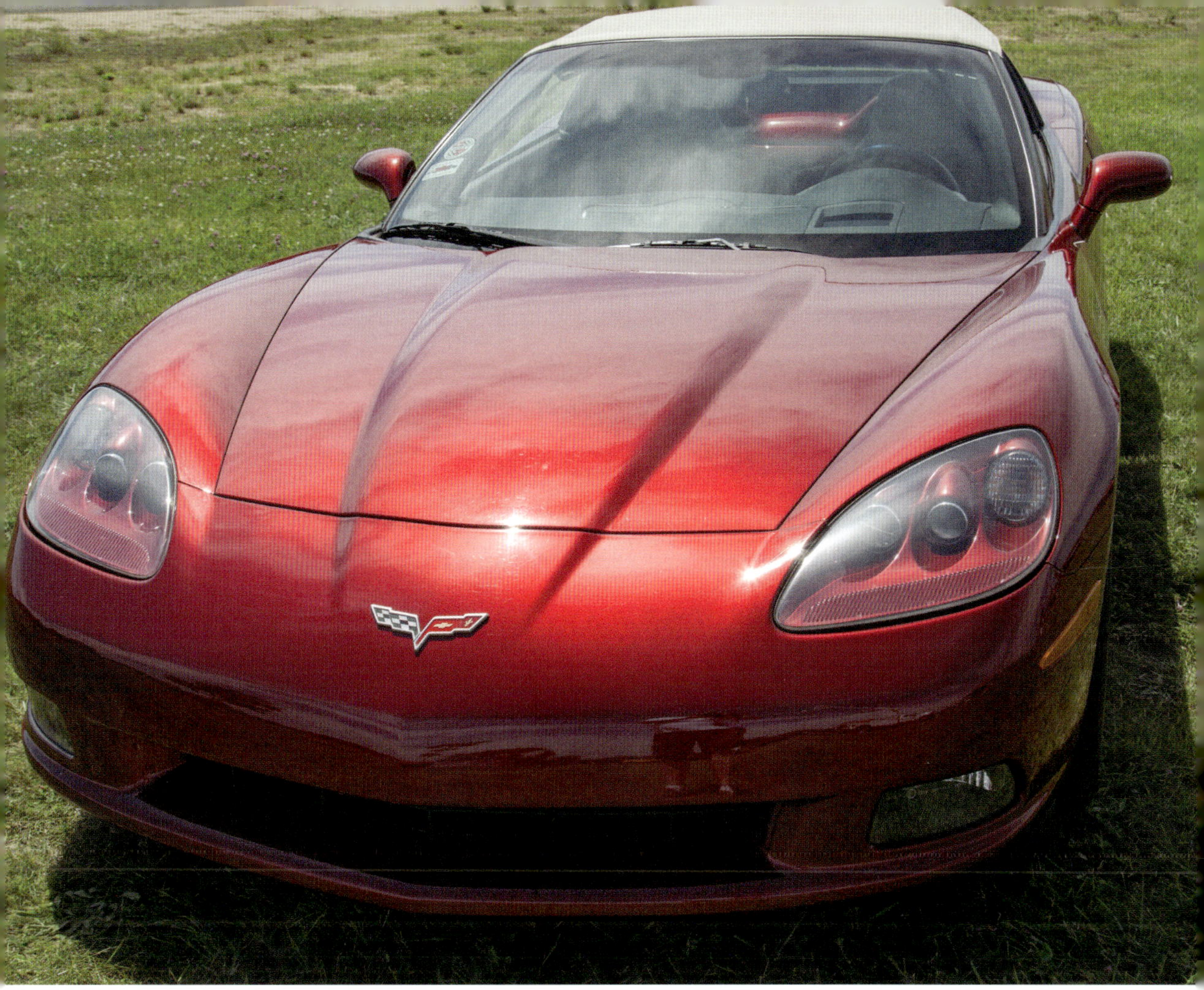

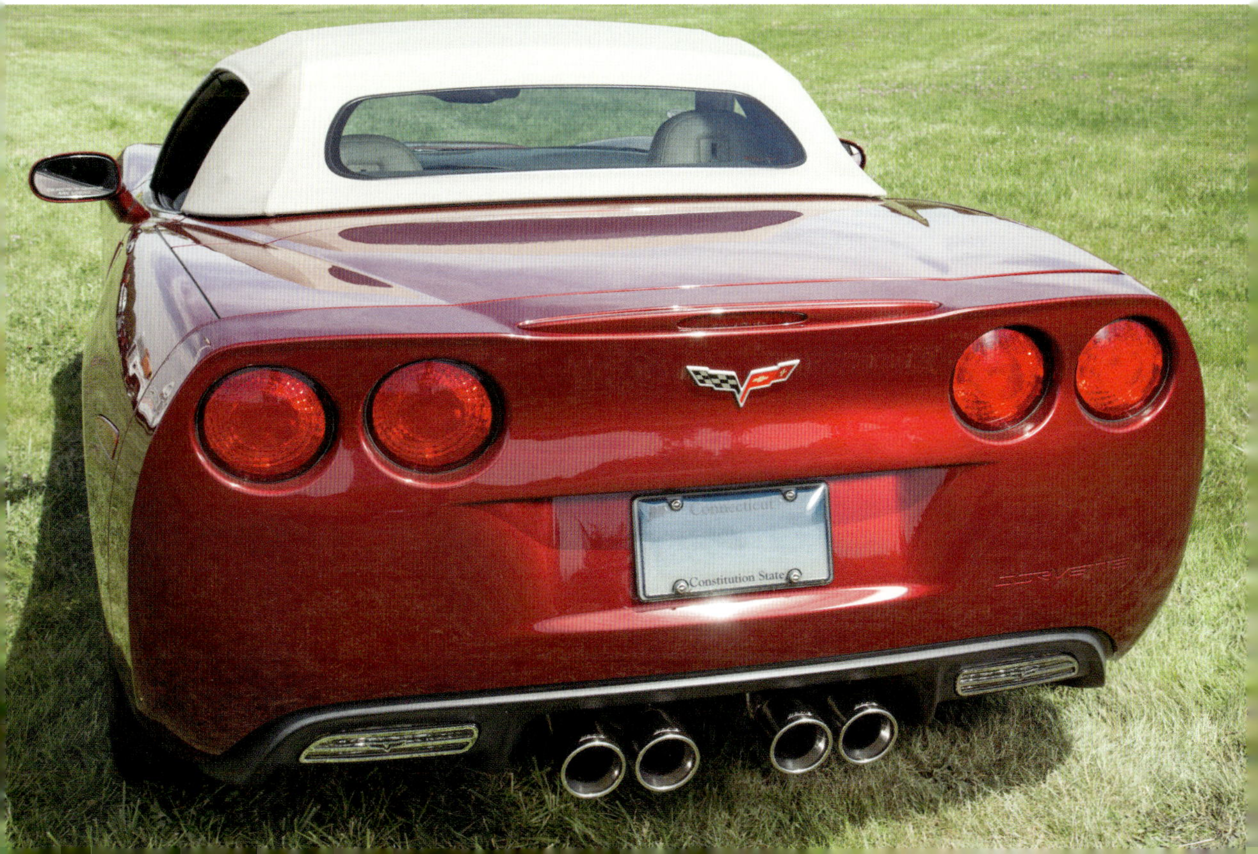
Connecticut
Constitution State

She Said Yes

1968 Convertible, Blue
Owner: Wayne

Wayne bought this blue 1968 Corvette in September of 1970, when he was single and 24 years old. Four years later, having enjoyed the Corvette as a single man for a few years, he decided it was time to settle down and marry his girlfriend.

"It was the Saturday before Christmas in 1974 and I planned to drive to her house and ask her to marry me. When I got in the car and hit the starter, all I heard was a grunt—the battery was too weak to start the car. Was this a bad omen?" he wondered.

But Wayne wasn't detered. "We lived on a hill, so I pushed the car out of my parents' carport, let it roll down the hill, and I cranked it with the clutch," he says. "I made it to her house, popped the question, and she said *yes*—leading to more than forty years of a terrific marriage!"

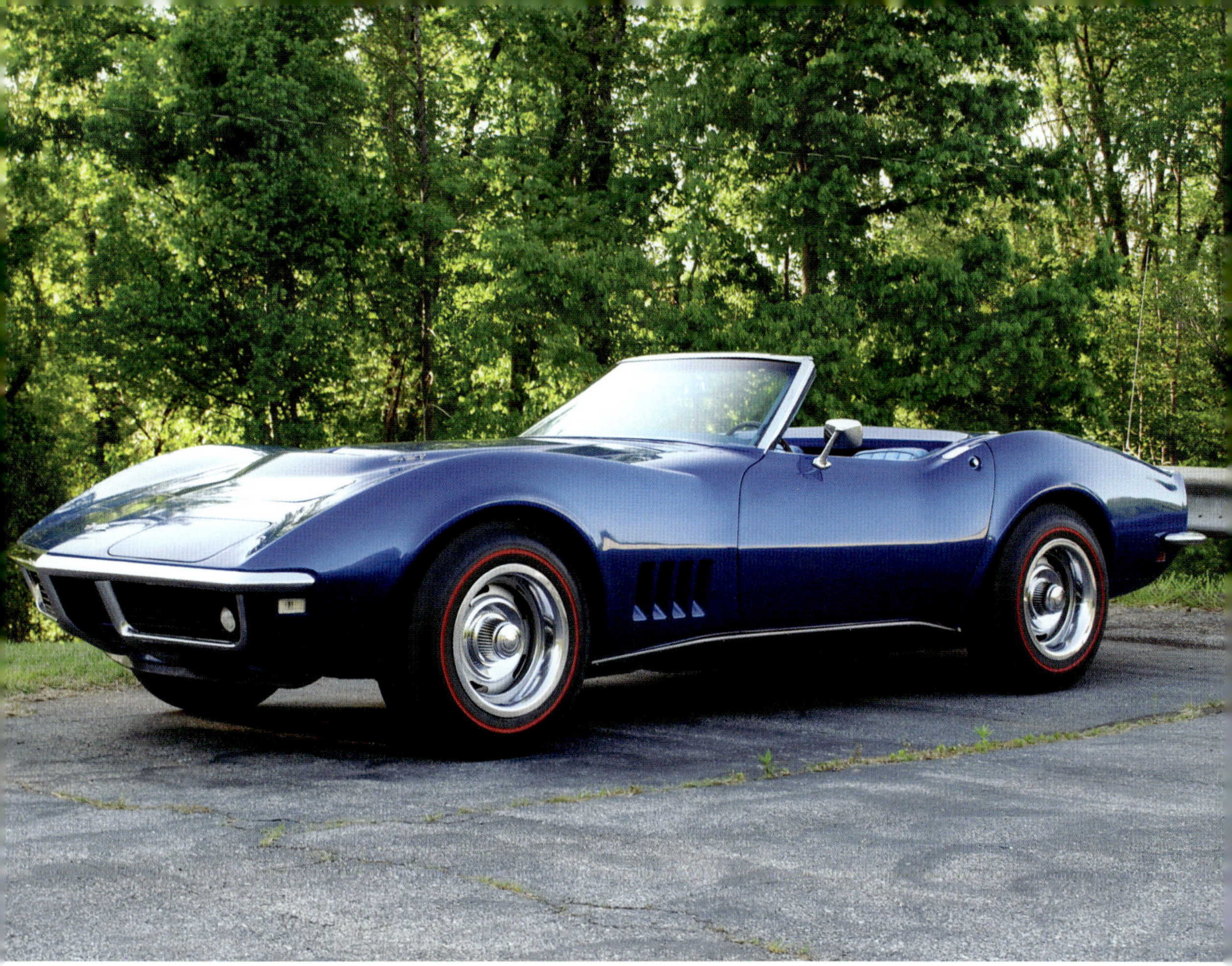

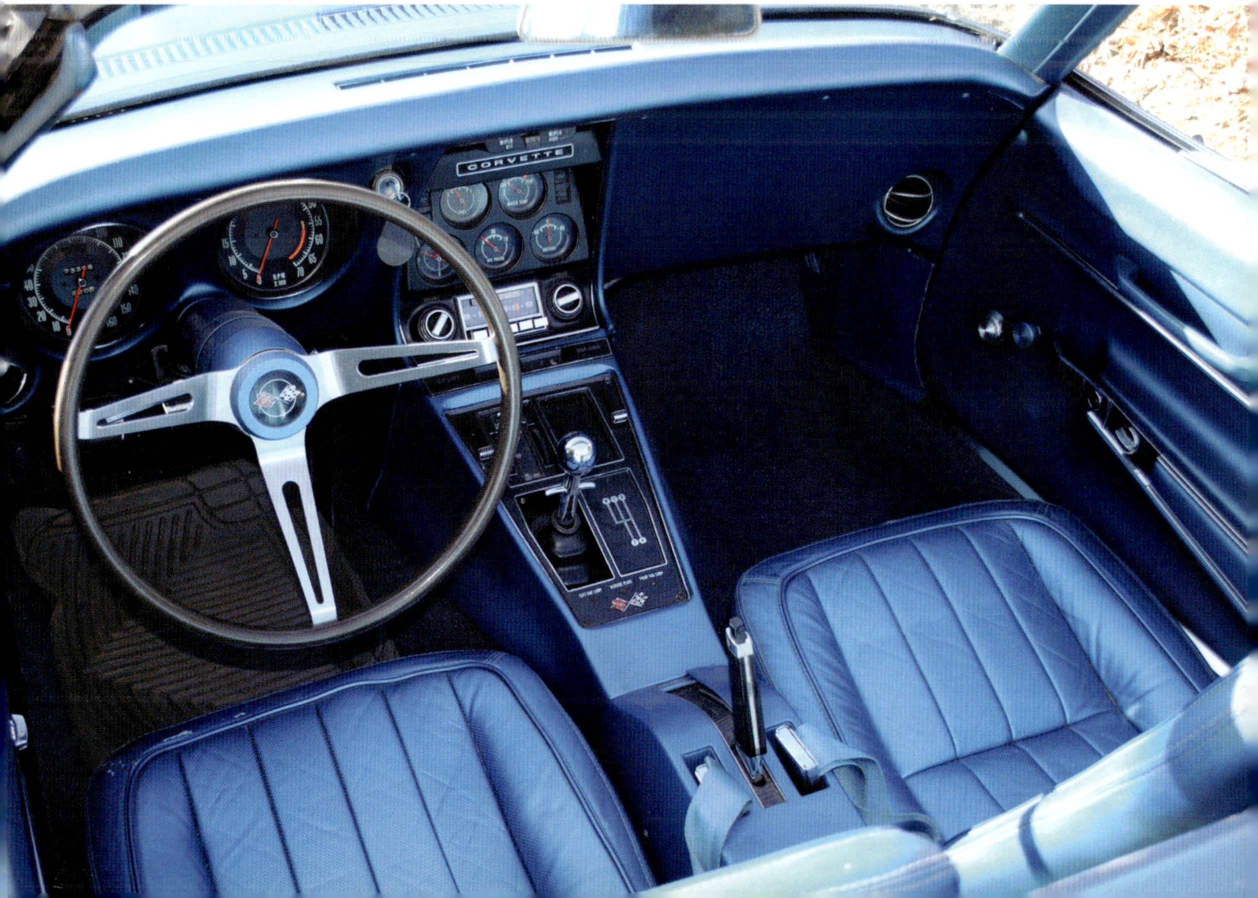
CORVETTE

Still a Thrill

1965 Coupe, Ermine White
Owner: Ray

Ray's 1965 Ermine White Corvette 327-365 four-speed coupe has a beautiful black interior. "Even though this machine is 53 years old, it still runs like new," he says. "I take care of this as well as my other slightly newer Corvette, a 1967 model. Being a Corvette owner and having the luxury and opportunity to drive them is a thrill I have enjoyed for forty years and never goes away."

A Promise to Himself

2003 Convertible, Red
Owner: Peter

Back in 1961, Peter's friend owned a 1960 Corvette. Unfortunately, his friend injured his right hand at work and was unable to drive because the car was a stick shift. "He made me his driver and we would cruise every night—especially in the summer," Peter remembers.

"After that experience, I said that I would own a Corvette someday. Many years later, my dream came true with this 2003 Red Corvette."

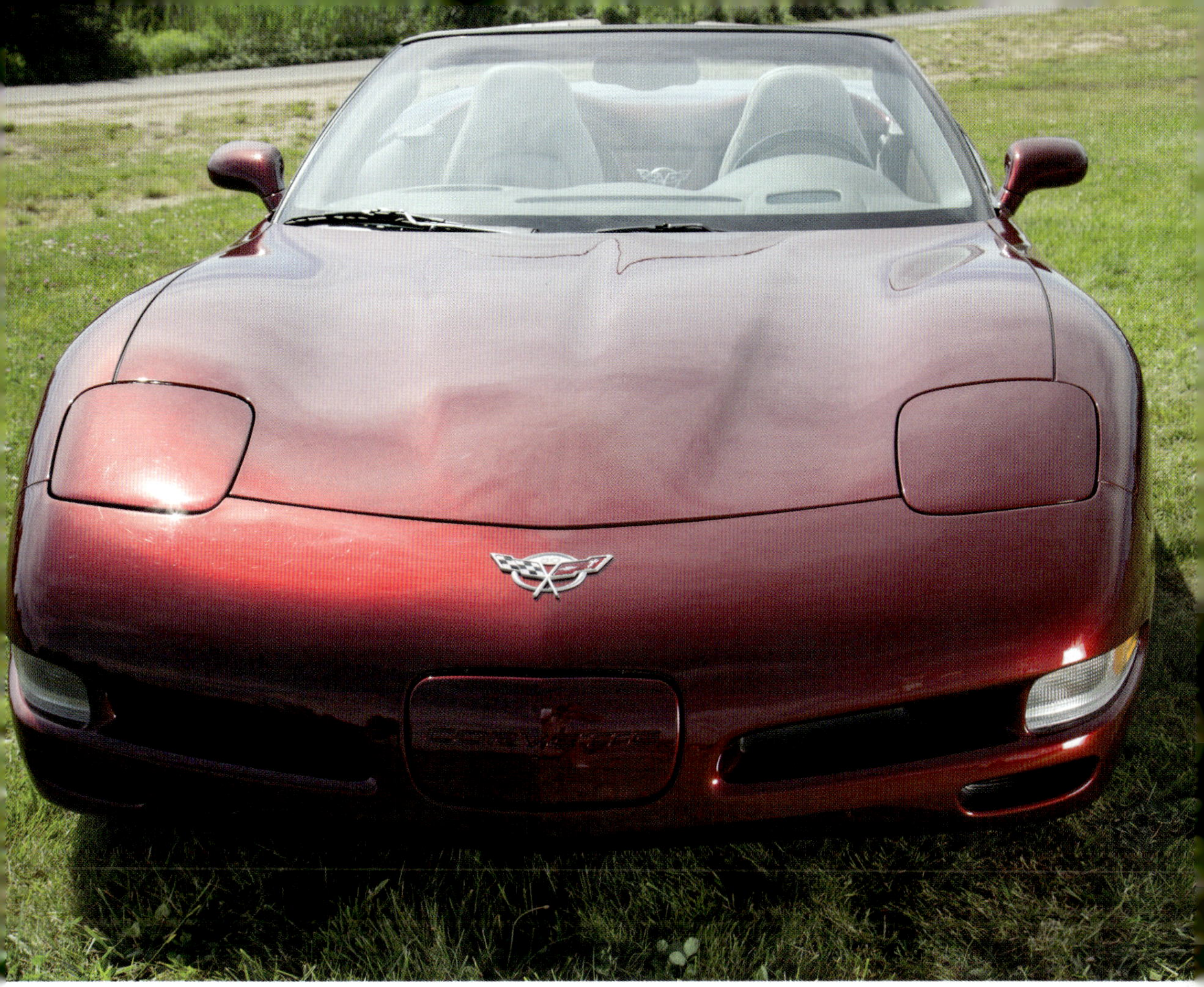

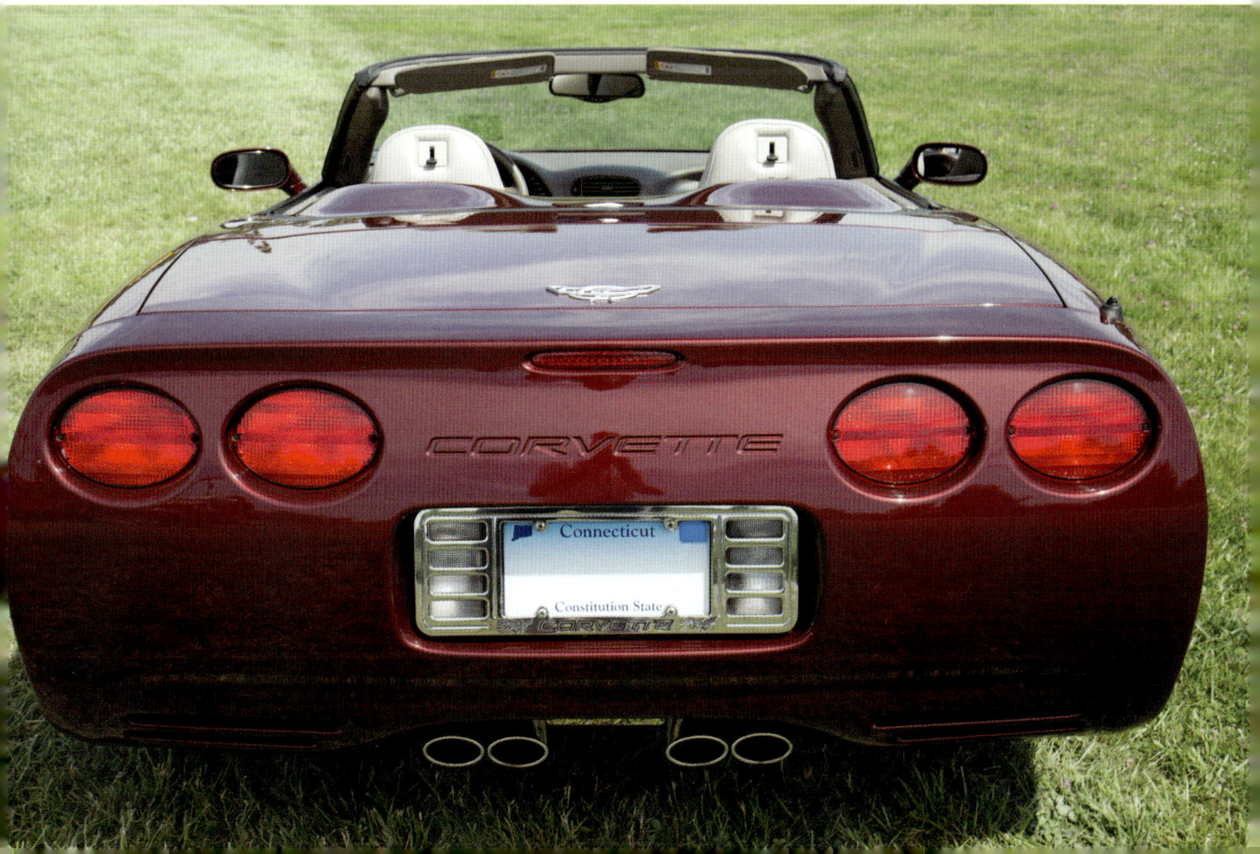
CORVETTE
Connecticut
Constitution State
CORVETTE

Little Red Corvette

2012 Grand Sport Convertible, Crystal Red Metallic

Owner: Harry

This Corvette is a dream come true. We all have dreams that keep us going from day-to-day, and Harry's was to own that "Little Red Corvette" made famous in the song. Unfortunately, the business of life often gets in the way and the dreams fade for many of us. At age 55, a serious illness brought Harry's dream of owning this car into sharp focus.

His dream is now five years fulfilled and the illness is resolved, but he still cannot believe that this 2012 Crystal Red Metallic Corvette Grand Sport convertible with a Cashmere top is his.

The moral of his story is pursue your dreams because life is fragile and dreams are important. Live your dreams!

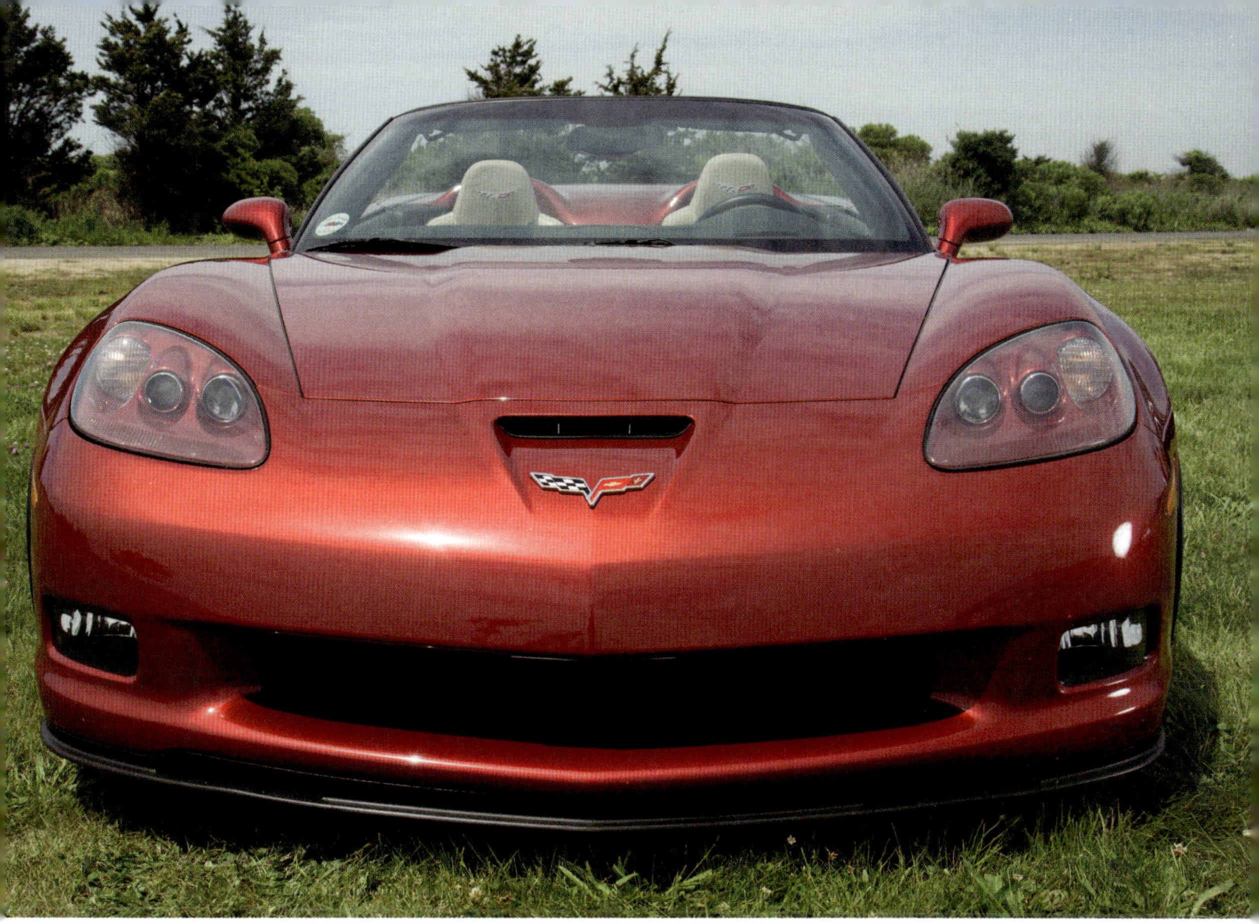

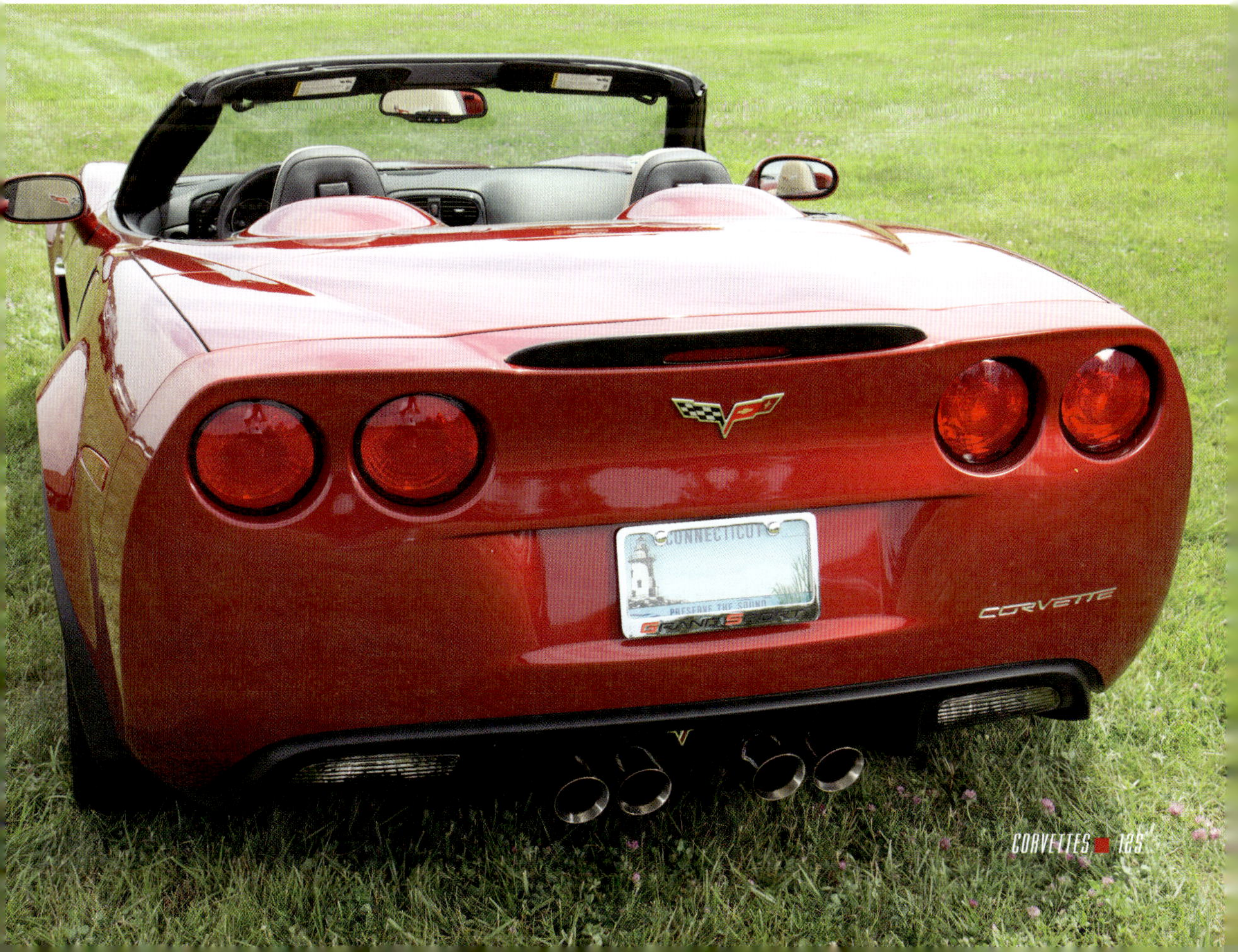
CORVETTE

Index

AmherstMedia.com

- New books every month
- Books on all photography subjects and specialties
- Learn from leading experts in every field
- Buy with Amazon (amazon.com), Barnes & Noble (barnesandnoble.com), and Indiebound (indiebound.com)
- Follow us on social media at: facebook.com/AmherstMediaInc, twitter.com/AmherstMedia, or www.instagram.com/amherstmediaphotobooks

Trees in Black & White
Follow acclaimed landscape photographer Tony Howell around the world in search of his favorite photo subject: beautiful trees of all shapes and sizes. *$24.95 list, 7x10, 128p, 180 images, index, order no. 2181.*

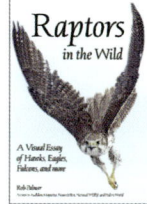

Raptors in the Wild
Rob Palmer shares his breathtaking images of hawks, eagles, falcons, and more—along with information on species and behaviors. *$24.95 list, 7x10, 128p, 200 color images, index, order no. 2191.*

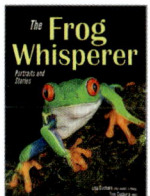

The Frog Whisperer
PORTRAITS AND STORIES
Tom and Lisa Cuchara's book features fun and captivating frog portraits that will delight amphibian lovers. *$24.95 list, 7x10, 128p, 350 color images, index, order no. 2185.*

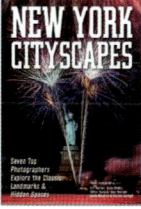

New York Cityscapes
SEVEN TOP PHOTOGRAPHERS EXPLORE
THE CLASSIC LANDMARKS & HIDDEN SPACES
Native New Yorkers explore the city's architecture and neighborhoods. *$24.95 list, 7x10, 128p, 200 color images, index, order no. 2201.*

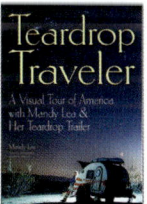

Teardrop Traveler
Photographer Mandy Lea left her job for life on the road—and along the way, captured both life lessons and incredible images of America. *$24.95 list, 7x10, 128p, 200 color images, index, order no. 2187.*

Birds In Flight
A VISUAL TOUR
Rob Palmer shares action-packed images of birds on the hunt, building nests, warding off invaders, and much more. *$24.95 list, 7x10, 128p, 200 color images, index, order no. 2198.*

Street Photography
Andrew "Fundy" Funderburg travels the globe and shows you how documenting your world can be a rewarding pastime and an important social tool. *$24.95 list, 7x10, 128p, 280 images, index, order no. 2188.*

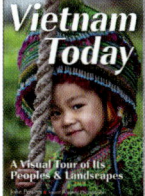

Vietnam Today
A VISUAL TOUR OF ITS PEOPLES & LANDSCAPES
Vietnam vet and acclaimed travel photographer John Powers explores the rural and urban sides of Vietnam today. *$24.95 list, 7x10, 128p, 180 color images, index, order no. 2207.*

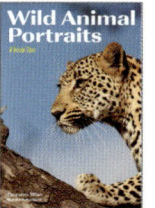

Wild Animal Portraits
Acclaimed wildlife photographer Thorsten Milse takes you on a world tour, sharing his favorite shots and the stories behind them. *$24.95 list, 7x10, 128p, 200 color images, index, order no. 2190.*

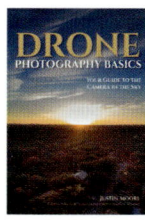

Drone Photography Basics
Justin Moore teaches you the fundamentals of using a drone and creating top-notch aerial photographs. *$29.95 list, 7x10, 128p, 130 color images, index, order no. 2211.*

AmherstMedia.com

- *New books every month*
- *Books on all photography subjects and specialties*
- *Learn from leading experts in every field*
- *Buy with Amazon (amazon.com), Barnes & Noble (barnesandnoble.com), and Indiebound (indiebound.com)*
- *Follow us on social media at: facebook.com/AmherstMediaInc, twitter.com/AmherstMedia, or www.instagram.com/amherstmediaphotobooks*

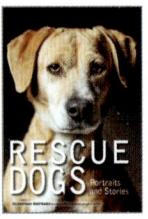

Rescue Dogs
PORTRAITS AND STORIES
Susannah Maynard shares heart-warming stories of pups who have found their forever homes. *$21.95 list, 7x10, 128p, 180 color images, index, order no. 2161.*

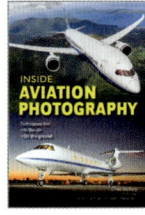

Inside Aviation Photography
Take wing with Chad Slattery as he photographs from the air and ground, shooting planes, pilots, and more. *$24.95 list, 7x10, 128p, 180 color images, index, order no. 2167.*

Create Fine Art Photography from Historic Places and Rusty Things
Tom and Lisa Cuchara take photo exploration to the next level! *$19.95 list, 7x10, 128p, 180 color images, index, order no. 2159.*

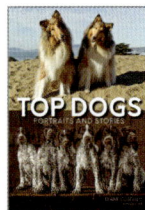

Top Dogs
PORTRAITS AND STORIES
Diane Costello shares stories about our constant companions and highlights the roles they play in our lives. *$24.95 list, 7x10, 128p, 180 color images, index, order no. 2169.*

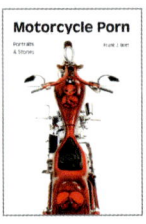

Motorcycle Porn
PORTRAITS AND STORIES
Frank J. Bott shows you the sexy side of these beautifully engineered and adorned machines. *$24.95 list, 7x10, 128p, 180 color images, index, order no. 2165.*

The Earth
NASA IMAGES FROM SPACE
Images from space reveal the startling beauty and incredible diversity of the blue orb we call home. *$24.95 list, 7x10, 128p, 180 color images, index, order no. 2170.*

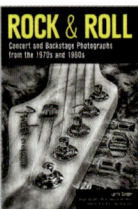

Rock & Roll CONCERT AND BACKSTAGE PHOTOGRAPHS FROM THE 1970S AND 1980S
Larry Singer shares his photos and stories from two decades behind the scenes at classic concerts. *$24.95 list, 7x10, 128p, 180 color images, index, order no. 2158.*

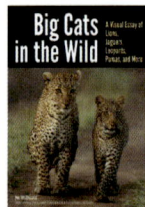

Big Cats in the Wild
Joe McDonald's book teaches you everything you want to know about the habits and habitats of the world's most powerful and majestic big cats. *$24.95 list, 7x10, 128p, 220 color images, index, order no. 2172.*

Hubble in Space
The Hubble Space Telescope launched in 1990 and has recorded some of the most detailed images of space ever captured. *$24.95 list, 7x10, 128p, 180 color images, index, order no. 2162.*

Owls in the Wild
A VISUAL ESSAY
Rob Palmer shares some of his favorite owl images, complete with interesting stories about these birds. *$24.95 list, 7x10, 128p, 180 color images, index, order no. 2178.*